Out Of Step

A Memoir of the Vietnam Years

Part One

by Francis Hamit

Contributors

John Reid & John Wadje

Edited by Leigh Strother-Vien

BRASS
CANNON
BOOKS

ISBN: KDP Paperback: 978-1-59595-082-6

Cover image by : **Markee Books**

Interior Design by : **Markee Books**

Brass Cannon logo was designed for us by **George Mattingly**

TABLE OF CONTENTS

DEDICATION

In memory of all those who died, and who are still dying, because they served in the Indochina theatre.

You are not forgotten.

OUT OF STEP:

A MEMOIR OF THE VIETNAM WAR YEARS

PART ONE

INTRODUCTION

"Our greatest shortcoming in Vietnam is the lack of timely actionable intelligence."

– A Special Forces officer

"If we see the enemy clearly, we will win. If the enemy sees us clearly, we will lose."

- Ho Chi Minh

With that simple statement in 1947, Ho Chi Minh set the rules of engagement that would guide the People's Republic of Vietnam's cryptographic services until final victory in 1975.

It was all one war for them, in a theatre of operations that included Laos and Cambodia as well as South Vietnam. Lines on a map drawn by others meant nothing to them. Starting with no cryptographic sources but an old manual in French, they built a world-class Signals Intelligence (SIGINT) establishment, in many ways superior to our own.

I was there. I witnessed many of the conflicting and highly charged cultural phenomena of the 1960s. I saw the first burning of a Draft Card, and the last anti-war riot in Iowa City.

This is part of a longer narrative. It covers my first two years in the US Army Security Agency and, more to the point, my one year tour in Vietnam, and what it means to be a Soldier.

This is the sequel to my Iowa City memoir ***"A Perfect Spy"***.

In the interests of clarity and accuracy, I have added brief passages posted on Facebook or sent to me by e-mail by other ASA soldiers who were where I was not.

A warning to parents and sensitive readers: there is a lot of sex here. It's also a coming-of-age story. Others who were there may remember some things differently. Memories are never entirely accurate.

When I enlisted in the US Army, I was 22 years old.

– Francis Hamit, October 2025

IN THE ARMY NOW

CHAPTER 1

Regardless of when they serve, everyone is in a different Army from everyone else. Every path through the service is unique to the individual who treads it. It is a time of discovery and adventure. I did not enlist in the best of circumstances, or for the most noble of motives.

In several critical respects, I was the author of my own misfortunes: by falling in love with the wrong woman, neglecting my academic studies, and allowing myself to be distracted by taking up undercover work against the sudden influx of drugs in Iowa City.

I could see the writing on the wall ahead. My academic probation would not be renewed, and I would be drafted – if I wasn't killed by a drug dealer first.

(For further details see my previous memoir "A Perfect Spy")

Unless I stepped up and enlisted, when every other guy I knew was actively trying to find a way to stay of the service, or, failing that, the Vietnam War, which had, by this time, gone from being a noble cause to 'a fate worse than death'.

My father was disgusted with my choices, and my enlistment in the US Army did not improve his opinion, because it was an enlistment. He wanted me to become an officer as he had. No one at the Iowa City Police Department or the FBI would acknowledge my role as an operative. They couldn't. Not just because it might compromise a future case, but another, ongoing, operation that I had no knowledge of. I didn't do it for money, or to get out of a drug charge myself, but because I was in a position to do it, and thought it my duty, as a citizen, to step up. I received no pay. That was why there was no record.

I told my recruiting Sergeant that I wanted a job in Military Intelligence, doing the same kind of work. Counterintelligence. He had something else in mind for me once he saw my test scores: the Army Security Agency.

He said it was the same thing, and either lied or was grossly misinformed. Actually, I doubt that he cared. He was a salesman with a quota to fill. Allegedly, Army recruiters at that time were encouraged to sell the 'no ASA in Vietnam' lie cited below.

The recruiter had me from the beginning. He said I would be doing something above and beyond the normal duties of a soldier. To his credit, it might have turned out that way had not Fate intervened.

The FBI did the security clearances back then. My undercover work would be verified. I would be in M.I. But first I had to do what every person in the military has to do: learn to soldier.

I had to wait two months to do that. I spent that time working out and jogging and getting rid of my personal possessions, starting with the trusty and never-fired Walther P-38 I had carried for my protection. I sold that to a local gun dealer. I entrusted a footlocker full of favorite books and record albums to a female acquaintance who would disappear with it entirely and leave no trace. My cameras and civilian clothes I shipped to my parents. My 1957 Ford Station Wagon went for ten dollars to a guy who used it for landfill on his farm. My writing I took with me in a big cardboard folder.

I went into the Army with a clean slate, at least as far as Basic Training went. I had no illusions about it, having grown up in the Army. It was the family business. I reported to Fort Des Moines with one bag and the thick white cardboard accordion file full of writing. From there, I, and a group of other men from the Iowa National Guard, who were about to do their six months active duty for training, were bused to Fort Leonard Wood, Missouri.

The first three days we were yelled at constantly, as we drew our uniforms and other gear and took a battery of tests. Not just the Army GT test, on which I had to get a minimum score of 115 to stay in ASA, but others that measured potential for specific skills. Radio code was one I got a very high score on, and my test for OCS mysteriously was scored just below the point at which I would have been placed on that track instead.

The explanation was on my orders. "Individual is under the Command and Control of the Commanding General, USASA and may not be reassigned." This, I was later told, meant that I wasn't actually in the Army at all, but something that looked very much like it. ASA, whatever it was, had me, and would not let me go to become cannon fodder for the war in Vietnam. ASA was a world-wide organization. The odds were much against me being assigned to the war zone.

The GT score, although I didn't know what it meant at the time, was life-changing. 147 out of 163. Like my high SAT score it meant I was a 'genius'. I didn't believe it and thought it must be a fluke of some kind, that I just tested well. I'd struggled with schoolwork all my life, and barely graduated high school. I had other problems that later generations would call 'dyslexia' and 'ADHD' and Asperger's Syndrome. The world wasn't there yet, even if the Army was. I was called stupid and lazy so much that I believed it. In the end, it just made me try harder.

I was endowed with a natural curiosity, and once I learned to read, I read all the books that people told me not to, just to see what they were about. What was being hidden?

But two courses in 'Remedial Reading' aside, in the 3rd and 8th grades, I got no special assistance. My education lacked continuity. Army families move every two to three years. Things taught in one grade were taught in another at the new post. I never learned to do a square root. Calculus is beyond my comprehension. My dyslexia is more 'dismathia.' Some mathematical concepts elude my mind and cannot be retained.

That's the wonderful thing about the US Military. They have a job for almost everyone. You may just have to stumble around a bit until you find your niche. Stumble I did.

At the end of the three days, we were transferred to various companies on post for eight weeks of Basic Training. Most of the company I was in were those same Iowa National Guardsmen I rode down with from Fort Des Moines. There was a sprinkling of draftees from Minnesota.

The National Guard had 'NG' as a suffix before their eight-digit serial numbers. The draftees had 'US'. There were four of us who had actually enlisted. Our suffix was 'RA', and we were all four-year enlistees rather than three, which meant that we were all actually members of ASA. It made us 'special' when we simply wanted to blend in.

"What the hell is the Army Security Agency?" asked my DI at morning formation the next day. "Private Hamit?"

"Military Intelligence, Sergeant!"

"Bullshit. Drop down and give me ten."

By which he meant ten push-ups. This was the punishment for any real or imagined infraction. It was part of the physical conditioning and also part of the ritual humiliation that all trainees were subjected to. I did so, and found out that I was not in as good a shape as I'd thought. Of course, the five-mile run in combat boots before breakfast might have weakened me somewhat. With difficulty I got it done. I stood up.

"Hamit, tell me again what the Army Security Agency is."

"Military Intelligence, Sergeant. At least so I was told."

"You were told wrong. Drop down and give me twenty."

That took a while. I got to my feet again.

"Hamit, what is the Army Security Agency?"

"I have no idea, Sergeant."

"That's right. We get some of you ASA bozos every cycle, and you all seem to think you're something special. That you all walk on water. Well, you're not in ASA until you finish Basic. So get that out of your mind. Got that, Hamit?"

"Yes, Sergeant! Loud and clear."

From postings for the Facebook pages for ASA veterans, I've learned that many ASA-bound Basic trainees experienced similar harassment, regardless of where they did their Basic training. It was part of the cultural divide that developed during the Vietnam War between the Regular Army and ASA. That lying recruiting poster made us 'special.' The DIs hated that. So did the draftees and the National Guard trainees.

Those ASA recruits who were avoiding service in Vietnam – and most of them managed that – hated the DIs and the Army for the extra attention. Where uniformity is the goal, being singled out is a special kind of continuing Hell. It created an 'us versus them' rhetoric among lower-ranked enlisted men. Career NCOs were denigrated as 'Lifers.' The anti-war movement took advantage of this growing divide to undermine the war effort in particular, and in the US military generally. The result was a continuing morale problem with effects ranging from what the British call 'dumb insolence' to men in the war zone killing officers by rolling hand grenades under their bunks.

We moved into brand new two-story barracks built with concrete and tile. We were in eight-man squad rooms rather than platoon-sized drafty, overheated wooden barracks of World War II vintage. Those I would experience at Fort Devens, and remember my father's comment that there was nothing so permanent in the military as a temporary building.

They took away my writing portfolio during the shakedown inspection the night before. "You won't have time for this," one of the sergeants said, not unkindly, promising that it would be returned at the end of the training cycle. Other trainees were also giving up things. Knives and handguns mostly, although one guy had a very nice bong. The sergeant who found it simply laughed and crushed it under his size twelve combat boot. If it was a ploy to get kicked out, it failed. But with potheads, who knew?

He may have thought it something permitted, like cigarettes. None of them were great thinkers.

There was one ASA soldier in each of the other platoons, and they were all singled out for the same treatment I got. Someone coming into ASA from a previous cycle had left an indelible impression on these DIs, and every ASA guy coming through D-3-3 would pay the price of his arrogance.

The major harassment came not from the DIs, but from the Iowa National Guard. They all knew each other and were mostly from Des Moines and Ames, some of them already college graduates with good jobs they didn't want to give up. Frat boys and country-club Republicans, who pitied those who were not smart nor nimble enough to find a safe place in the Guard. A place so safe that they were never going to be bothered by participating in that nasty war in Southeast Asia. After Basic, they were all going to learn how to drive heavy trucks – they were already assigned to a transportation unit. In the unlikely event of a call-up, they would go to Europe. They had it all figured out, and were especially critical of RAs. The draftees couldn't help themselves, but we had been stupid enough to volunteer.

But there is no ASA in Vietnam, one of the other ASA soldiers said.

"Yeah, right!" Even the Iowa National Guard knew that for a lie. They had the political connections to find the really safe assignments. To a man, they were for fighting the war, but somehow too special to do it themselves. All the sons of rich, powerful men with connections.

One of them later told me privately that his father had threatened to disinherit him, if he actually volunteered for the Regular Army. Their business needed him more. The rich are different from you and me.

I hated these guys for decades after this, until the Iraq War, when that same unit was called up and drove four million miles in combat conditions. Some of them were killed in action. Then I was finally able to forgive them and put that grievance to rest. The sons repaid the psychic debt incurred by their fathers and grandfathers during Vietnam. They lost people and bled a little.

Back during our war, the National Guard was a certified refuge from combat zone service. The Army changed the rules by the 1980s so this couldn't happen again, and went to an all-volunteer force. Career military officers were the first to advocate that the Draft be abolished.

All of the trainee leaders were selected from the National Guard. The Drill Instructors and Assistant Drill Instructors were likewise very young men, all of draft age.

It was not until I moved to Pine Mountain Club in 2002, and met the local newspaper owner, Kim Knoller, that I found out why. He was a drill instructor at Fort Ord, California. He was offered the job when he was drafted, with the assurance that he would never be sent to Vietnam. He was still embarrassed about this 35 years later, but that was how the Army met a sudden demand for Drill Instructors – by offering draftees a chance to abuse their betters and an exemption from service in the war zone. The sudden war-time expansion of trainers could not be filled by career NCOs. This was a quick and easy solution that satisfied no one.

The job attracted a lot of bullies, and bullies encouraged bullies within the ranks. In our Second Platoon (I was in the First), the platoon leaders took to taking members of their platoon who had not measured up into the bathroom after midnight and giving them 'correction,' i.e., beating them up. They went too far with one victim, who reported for sick call and told the E-6 in charge of the local dispensary. He immediately reported the assault and battery to the MPs.

The other DIs and the company commander, a First Lieutenant, tried to cover this up, but the DIs in charge of that platoon, who'd encouraged this nonsense, both got Article 15s, as did their hapless company commander, from the Battalion's commanding officer, and the trainee leaders were given Summary Courts Martial.

At this point we received, by levy, a real DI who'd been a career NCO for sixteen years. His name, which I shall always honor, was Benny Bowman. He knew what he was doing and then some. In appearance, he resembled the actor Glenn Ford and had that same calm presence. He was senior in rank to all the other DIs, and his real job was to train them, as they tried to train us. He had spent the previous five years on the Army Pistol Team in West Germany competing in various marksmanship events, but he knew what being an NCO was about.

Our marksmanship training was on the M14. This was a weapon developed by the Army to replace the M1 rifle that was its primary weapon during World War II and Korea. It was designed for war in Europe and took the 7.62 mm NATO round. The Army was already starting to replace it with the M16, a much lighter .223 caliber rifle with a plastic stock developed by the Armalite Corporation and manufactured by Colt Industries. It was designed for jungle warfare.

The M14 is heavy, about nine pounds empty. Carrying it all day at port arms or marching with it on your shoulder in close order drill is a challenge in itself. I expected to do well, since I was part of the gun culture in high school and went to the range frequently.

The rifle I was issued, however, was worn out and the sights problematical. The rear sight moved every time it was fired. That put me at a disadvantage, especially on the thousand meter range. My primary DI, a tall, lean, hard-drinking former construction worker who relied on verbal abuse as his primary motivational tool, was not very helpful, and refused to get me another weapon that had sights that would stay locked in place. I was reduced to using 'Kentucky windage'.

I was shooting for my qualification and he was shouting threats in my ear, as if that would magically cure the piece of shit M14 I had been issued, when SFC Benny Bowman said, very mildly, "Sergeant. Do you think it's a good idea to be yelling in that man's ear when he's holding a loaded weapon?"

The young DI, startled, shut up, and I qualified at the lowest grade, Marksman. I thanked SFC Bowman, and asked for another shot at a better score. He looked at my weapon and shook his head. "Not with this. I'm surprised you did that well, and we have a schedule to keep. Look at it this way. You won't be recycled."

The hard reality is that I was not an outstanding trainee. I had problems with marching, and usually led off on the wrong foot. That threw off other men marching with me. One big farmer from Minnesota took to marching behind me and deliberately stepping on my right heel. He didn't apologize, but made it plain he had it in for me. His repeatedly stepping on my heel finally fractured it, but I never pointed the blame toward him. I figured I wouldn't be seeing him again after the cycle was complete. He wasn't the problem anyway. The DIs and ADIs were.

"You read too much," was his explanation for his bullying. Complaining about this was only likely to earn me more push-ups. The scandal of the beatings in Second Platoon had not yet broken. The right heel he injured troubles me to this day. That is compounded by the injury to my right knee. I'll get to that in a bit.

CHAPTER 2

My father wrote me:

> *I am truly sorry you are having a bad time but I know and you should have known that it would be no picnic. There is nothing I can do to make it any easier, and I don't think you would want me to, if I could. At least I hope not.*
>
> *You are in no position to be antagonizing anyone... the sooner you get down to business and do what they want you to do, the better you will get along. Physical training counts a lot and if you think the PT in Basic is rough, wait until you get to OCS... I don't mean to be unkind, and wish there were some way I could run some interference for you, but there isn't.... You have finally gotten into a position where you must fish or cut bait; there is no other choice – except to be a fink, and I know you don't want that.*
>
> *So pull yourself together Old Buddy and show them you can do it and have got what it takes. When you do that, I am sure things will go a lot better and easier.*

Of course it was a mistake to tell him anything about what was really going on, and, as he usually did, he had minimized and distorted what I told him to maintain the mental image he had of me as an irresponsible child. We were all required to write our parents every week to keep them off the Army's back with 'what have you done with my child' phone calls and letters. And what else could I have written about? The class-consciousness of the Regular Army showed in that letter, too, with its plea to become a commissioned officer as soon as possible and stop embarrassing him.

There was nothing unique about the rest of the cycle, but the entire thing was an indicator of how overstretched the entire Army was by the demands of the war in Vietnam. The Draft, which swept up all kinds of inappropriate people in its maw, was part of the problem. The young DIs and ADIs had been taught that anyone could be made into a soldier with enough training and 'motivation'. Hence all the yelling and verbal abuse.

But some of those drafted were unfit, and never should have made it as far as Basic Training.

One was an older man. The draft cut off at age 38. He claimed to be 42 but had no birth certificate to prove it. He claimed to be married, with four kids, another fact which should have made him exempt, but the marriage was common-law and undocumented, and his neighbors on the local draft board expressed their moral disapproval by sending him a notice which he could not get out of.

He was in Third Platoon, and the guy who fell out first on the morning run. We were actively discouraged from going on Sick Call, and accused of malingering if we did. His complaints about chest pains were ignored. Shortly before Benny Bowman joined us, he had a mild heart attack, was taken away in an ambulance, and never seen again. They'd come close to killing him because he lacked the proper papers.

I became ill myself. The young DIs had the notion that our food had to be restricted and that going without water while we trained would make us tougher. This was counter to official doctrine. That encouraged eating as much as you could. Ice cream was served at breakfast along with eggs, pancakes, bacon and sausage. Other meals were loaded with potatoes, vegetables, and desserts. All forbidden by our amateur trainers. Their model was out of Hollywood, and the Jack Webb film, The D.I. They quoted lines from it. They showed it to us, too, as a training film. I had a dramatic affect on me My Army medical record from that month describes me as "Anorexic" in big red letters.

Another film, more recently created by the Army for training purposes, starred Tony Dow, the erstwhile star of the television series Leave It To Beaver. Dow had been drafted and, like Elvis before him, had stepped up to do his duty. He was cast as the young naive trainee opposite an as-yet-unknown Canadian actor, William Shatner, who played the tough but kindly DI. Mr. Shatner's performance was very broad, so much so that one voice piped up in the 900-man audience watching it in an Army movie theater with, "Boy, he sure likes himself, don't he?"

The room exploded with laughter and the lesson was lost. Shatner's DI didn't come close to the amateurs imposed on us, but he was no Benny Bowman either. Bowman was a straight talker and quietly modest about himself and his career. In Shatner's defense, it should be noted he didn't write those heavy-handed bits of propaganda he was spouting. But not for nothing is he known as 'The Canadian Ham'. His performance was a bit of sly parody.

The other draftee who was a tragic figure was a guy who should have never made it past the induction station. He was a moron. Literally. A danger to himself and others. A product of Secretary Robert MacNamara's scheme to expand the draft by taking 100,000 men with low IQs who would normally not be allowed to enlist. There are jobs in the military that low IQ personnel can do, but they are hard to train and hard to manage. This teenager was an extreme case.

The term used to describe his condition today is 'developmentally disabled'. He needed constant supervision, and why he was not simply removed from the company and sent home is a mystery. He had to be told how to do even the simplest things, and otherwise would sit or stand until prompted, staring blankly into space. His presence was not, as with the older man, a case of local draft board malice, but of extreme negligence, compounded by the fact that the war inspired a lot of schemes to avoid military service. Some of them were very creative, and the standard ruse of confessing one's supposed homosexuality or showing up for the physical in women's clothing no longer worked. The bored reply from the audience at the reception centers was usually, "We don't care. You're still going," or "Pass on, Private Pussycat."

They probably thought he was faking it. No one could be that stupid. But he was.

He was given to me to supervise one day after I was placed on 'light duty.' The job was for him to cut the grass around the battalion headquarters, a newly-built yellow brick office building next to a condominium of company headquarters offices and supply rooms. The barracks were, as previously mentioned, rows of new two-story barracks that replaced the ones built during World War II. The war provided money for the Army to undertake some long-delayed capital improvement projects. The base hospital was a new multi-story building.

My supervision of this young man, I regret to say, consisted of me telling him to drive the riding lawnmower he was on in a particular direction and then to stop when he got to the end of the lawn, in increasing louder and more strident commands laced with profanity. It was extremely frustrating, and the job would have been done quicker had I simply told him to sit while I drove the lawn mower. But my light duty status forbade that, and there was the very real danger that he would wander off into the traffic and be injured. So I continued to try and 'supervise'. The draftee DIs had a good laugh over that.

Suddenly, there was an African-American Lt. Colonel standing next to the battalion headquarters building, holding a file of papers in his hand, without his jacket or cover. He looked more like a preacher than an army officer, and spoke very softly. He was the Battalion Commander. I was in trouble.

"Private. What are you doing?" he asked, very mildly.

I turned and saluted, feeling my face flush red. "Sir. Permission to speak freely?"

He nodded, now looking at the kid on the riding lawnmower who was staring, vacant-eyed, at both of us.

"This man should not be in the Army, sir. He's stupid."

"We have lots of stupid people in the Army, Private. They get by."

"Yes, sir. This man has not been able to do that. He requires constant supervision."

He nodded gravely. "Well, stop what you are doing. It's an embarrassment. Both of you go back to whomever assigned you to this detail, and tell them to report to me ASAP. And to bring your company commander."

I saluted him again. "Yes, sir. Thank you, sir."

A day later the young man was gone. The company commander and the malicious prick of a DI who'd devised that prank were in deep shit. This was the same guy who'd encouraged the frat boy hazing of trainees by their platoon leaders.

With the exception of SFC Bowman, the DIs and ADIs simply didn't know how to train the men properly, and were replicating their own recent experiences of abuse at the hands of men much like them. It was a vicious cycle repeating itself. Cruelty was part of it. What was the harm of a little fun?

Older, more experienced, NCOs in the program were assigned as instructors on various 'ranges'. Most of them were bitterly awaiting the time they could retire from the Army at 30 years of service and a full pension, and were on their last tour.

This war was not for them. They'd served in World War II and Korea, and they'd done enough.

As said above, I wrote my father about these problems, and he wrote back that the Army had a lot of experience in turning young men into soldiers. I should not judge my company commander too harshly, that his was a difficult job.

There was a myth that depriving trainees of water for long periods would toughen them up. No one knows for sure what causes gallstones, but there is some suspicion that dehydration might be a factor. Especially in a patient who is an otherwise healthy male of 22 years. I began to have abdominal pains, especially during the many runs that were required of us. Five miles in new ill-fitting combat boots before breakfast was just the start. We double-timed everywhere. To pass out of the Basic Training course, I had to run a mile in combat boots in 15 minutes. We practiced that several times a day between lessons where we sat and learned about first aid, various weapons, tactics, and military law and courtesy. Sometimes we were shown training films. These intervals were a welcome relief. They allowed you to catch your breath.

I'd thought of myself as pretty tough, but I was having major problems keeping up. I had a few friends. These were the guys who would help you through a problem. One of them was The Preacher, an older guy with a degree in Divinity who was part of the National Guard contingent, but not from that self-selected group of elitists just doing their time while they got on with their careers. He was quietly up for the fight, believing the enemy to be 'godless Communists' and simply trying to find the best place to serve. For some reason, he didn't care to become a Chaplain. His Christianity was of a more robust, militant character. I was a godless agnostic, but we liked each other, and got along because we never discussed religion.

It was he, seeing how sick I was becoming, who urged me to ignore the harassment from the DIs and go on sick call. I did, and the local medic, an E-6 who'd served in Vietnam and despised the DIs who prevented trainees from going on sick call, cursed them openly as he took blood and urine samples and called the hospital to get me a follow-up appointment.

I think The Preacher had a quiet word with SFC Bowman, and I was allowed to go over the objections of the DI who had charge of my platoon. He had marked me down as a malingerer and poor soldier because of my marching problem.

The next day I passed out on a one-mile run. I remember the DI, now in a panic, standing over me and yelling my name, trying to get me to stand up or just respond. I was laying face down in the dirt, and thought to myself, 'No, thank you. I'm just fine right here.' There was a sharp pain in my abdomen right under my rib cage and I wondered if one of the drug dealers from Iowa City had sent someone to take another shot at me. It hurt like hell.

An ambulance was called. I was taken to the hospital for further examination. The X-rays revealed a large gallstone. This was so unusual that several doctors came to look and to examine me. They had never seen the like in anyone so young.

"That's going to have to come out," one of them said. "You'll have to have an operation." He looked at me curiously, "By the way, are you Colonel Hamit's son?"

Reluctantly, I admitted this was so. The Army Medical Corps was like a small town. Everyone knew everyone, at least by reputation. And my father was modestly famous, on his fifth and last year as Chief of Surgery at Brooke Army Medical Center. Everyone knew his name. He'd done great things in medicine and was a highly skilled surgeon. The CPR course I took in Basic had been developed on an Army Medical Corps contract he'd administered ten years before. He helped Basil Pruitt to create the famous Army Burn Unit at Brooke. Colonel Pruitt was one of his Residents.

Hal Hamit was overbearing and used to getting his way. The doctors and surgeons at Fort Leonard Wood Army Hospital were intimidated. He'd already contacted someone there with a casual 'look out for my dumb kid' phone call before I got sick. Everyone has friends. Friends who will bend and even break the regulations.

I got a letter from him telling me that he had discussed my case with the Hospital Commanding Officer. They agreed it would be better handled at Brooke Army Medical Center in San Antonio. This was extraordinary, because it wasn't even of the same part of the Army Group as my Basic Training Company. He was nowhere near my chain-of-command, and I liked it that way. And, of course, I had not been consulted. He still thought of me as a child needing his protection and guidance. I strongly disagreed. This interference was diametrically opposed to what he'd written a few weeks before. Simply put, he knew better.

Despite the age of the patient, there was nothing extraordinary about the operation itself. My primary doctor, Captain Reynolds, was apologetic. "Your father thinks it would better if you were transferred to Brooke for the operation."

"Can he do that?"

Reynolds nodded. "No one here is going to argue with him." Of course not. Reynolds was a Captain and a career officer. Dad was a Colonel and might be a Brigadier General soon, in a part of the Army where flag rank officers could be counted on the fingers of one hand.

I met the surgeon who would do the operation. He was a very sharp, highly intelligent African-American with a brisk attitude who carefully explained the procedure. I could imagine my dear, sweet, but incurably racist, mother's reaction when she found out that a Black man would be the cutter on her only son.

But Captain Reddy C. Lemons, MD, FACS, showed me his hands when I asked. They were small, strong and well formed. I asked him to do that trick I had seen Dad do so many times, tying a square knot in a cat-gut suture with his thumb and forefinger. He looked at me curiously. Where had I learned about that? Rather than tell him who my father was (something I suspected he already knew), I just smiled.

"I'm in theatre," I said. "It's an audition."

He laughed. It took him less than a second to demonstrate his skill.

"This isn't going to happen," I said to Reynolds. "I will make a phone call and fix it."

"Well, we can schedule you for next week."

"That isn't going to happen either. I have two and a half weeks to go in Basic. I am not going through that again."

Reynolds and Lemons looked at each other.

"I'm in ASA. I'm scheduled for a school."

"You wait to have the operation, and you'll probably miss the school. It will take you some time to heal."

"I get recycled and have to start over, then I miss it anyway. ASA can drop me, and I'll end up in the Regular Army as a grunt. I didn't enlist for four years for that. I will finish Basic unless there is a reason you think the operation has to be done right now."

"No reason if you can gut it out. You're going to be on light duty and in a lot pain. You'd better talk to ASA."

The clerk at the ASA Personnel detachment was considerably put out when I told him of my problem. "You are scheduled for a class at the Defense Language Institute in Monterey. It's eleven months. You miss it, and I'll have to find something else for you."

"What language?" I was bemused by this, since I had done so poorly in languages in high school.

"Farsi."

I laughed. "What is that?"

"Persian. You'll eventually end up in Iran if you pass. Can't this operation wait?"

"I'm having a hard time convincing them to let me finish Basic. My father is trying to pull me down to Brooke for the operation, which adds more time."

"Who is your father?"

When I told him, he was, to my relief, less than impressed. "You're in ASA, Hamit, not the Army. You can't be reassigned unless the Commanding General of ASA says so. For anything." He paused and asked, "What do you want to do?"

"The last thing I need is one of my father's ham-handed residents cutting on me with him looking over their shoulder. Keep me here, please."

"I will. Call me when you're recovered, and I'll find another school for you."

When I called Dad, I got right to the point. "What are you trying to do, Colonel? Ruin my career before it's even begun?" I was pissed and had every right to be. I was 22 years old, and no longer a child, and he damn well knew it. And ASA was also a small town where everyone knew everyone. This incident would be talked about on the back channels. He sent me a letter a few days later. He was not convinced I was sincere about staying at Fort Leonard Wood.

"I called Captain Reynolds this morning to talk to him about your medical problem, and about the other problems that moving you will generate. He can arrange to have you evacuated to Brooke on the air evac plane as soon as you finish Basic Training. The point is that you would not have to lie around the Leonard Wood hospital. Of course if you would rather have the job done there, I am sure that Captain Reynolds and his staff are perfectly competent.

Of course they were, and he damn well knew that, too. Again, I sensed my mother behind it all, because he knew better. This was unacceptable 'command influence' even in the old Army he was in.

Perhaps, now that I was out if danger, he'd finally gotten around to telling her about the real reason I'd left Iowa City. He was about to transfer to Washington, and perhaps I would like to use my convalescent leave to go with them? (Not bloody likely. Traveling with him was itself a bit of an ordeal, but I couldn't say that.) There would be a lot of questions about Iowa City that I wasn't prepared to answer. Mostly having to do with my choice of female companions and my predilection for dating Jewish girls. I had learned, as do most people that age do, that confiding in family was likely to bring more problems than it solved. There was nothing heroic about any of it. And I was still deeply in love with the little bitch who'd almost gotten me killed. I had her photo in my writing portfolio.

First, I had to finish the cycle. The light duty designation meant that I still had to go to training. My DI said, "It seems you're some kind of weakling."

I stared at him. He just didn't get it. His training methods might have been the cause of my disease. I shook it off. As my father said many times, arguing with fools is a waste of time. I turned to go.

"How did someone like you end up in Military Intelligence, anyway?"

I gave in to temptation. "It might have had something to do with what I really did in college."

"Drama major? You were a faggot?"

I ignored the provocation. "No. I was an undercover narcotics agent."

It was his turn to stare. "You're shitting me. You couldn't do that on a bet."

"That's where the acting training came in handy, Sarge. I was playing a role."

I left him to chew that one over. He eased off a bit after that. But the fire and maneuver range would make me the stuff of minor legend. The exercise was to simulate an assault on an enemy position. Just like playing war when you were a kid. Except that I fell and hit my right knee on a large rock. By the following morning it was swollen to the size of a large grapefruit and filled with pus. The gallbladder disease lowered my resistance to infection.

So it was back to the hospital where I was issued antibiotics and crutches. I had a week to go, and I would damn well finish Basic. There was some resistance on the part of the DIs, but finally it was agreed. Some exercises like the low crawl, I could do, and those forbidden to me because I was on light duty, I could observe others doing, and comment on what they did right or wrong. So I became famous as the trainee who finished

Basic on crutches, held up to subsequent cycles as an example of determination and guts. I don't know about the guts part, but I was very determined to not repeat Basic. The Army had other things in mind for me. What, exactly, was not clear, but better than anything on offer at Fort Leonard Wood.

My Basic Training company First Sergeant was suddenly very impressed with me. The orders for Language School had arrived along with everyone else's. The National Guard guys were going to the Transportation Corps school to become heavy truck drivers. The draftees were going to Infantry AIT and, from there, likely, to the war in Vietnam. No one was gloating about this by that time.

"Language school," the First Sergeant said; "You must be a very smart man." He was a Puerto-Rican in his 40s who spoke with a heavy accent, and who never went to the field where he might have to take note of how the DIs and ADIs did their jobs. He was drowning in paperwork and didn't have a company clerk.

"You could stay right here after you get out of the hospital," he said. "You wouldn't have to go to Vietnam." Bodysnatching trainees to become clerks was another 'field improvisation' for the rapidly expanding training brigade at Fort Leonard Wood. The idea appalled me.

"Frankly, First Sergeant, Vietnam doesn't scare me. Staying here for four years? That terrifies me, because I'll die of boredom. But you know there is no ASA in Vietnam, and neither of us can get my assignment changed. So thanks, but no thanks. I'll dance with the fella I came with."

He understood about half of this, and would bring the matter up several more times before I finished my recovery and went on to Fort Devens. It was not that he liked me so much as that he needed a smart clerk. Those were in short supply everywhere in the Army.

This was a turning point in my life. The language school would have meant being assigned to Iran and working a different kind of assignment than most ASA soldiers got. And never going to Vietnam or being poisoned by my own government with Agent Orange. I would read that detachment's Annual Historical Report with some envy a few years later. It was close to the Hollywood and popular culture version of what being a spy was about. That detachment wore civilian clothes provided by the government, lived in a fancy villa, and drove or were driven in unmarked Army cars. They had servants. An incredible assignment. But it was not to be.

CHAPTER 3

The orderlies on the surgical ward, both male and female, were particularly friendly to me. They had heard about how I had stood up to my old man and refused to be pulled away from their care. They were all enlisted volunteers in the Medical Service Corps (MSC), the part of the Army that my father saw daily, and he had threatened to disown me if I ever joined.

The reason became clear. Some of them were openly homosexual. One light-skinned African-American female E-4 admired my former fiancee's picture so much she asked for her phone number.

"I'm bi," she said. Not something you normally heard in the Army at that time on first acquaintance. It was actually a violation of the UCMJ if she were caught in the act. But she said it as if it were the most normal thing in the world, and maybe it was in the MSC, but the rest of the Army and the rest of the world wasn't there yet. Things were changing, but not that fast.

As I was checking in and getting installed in the bed I would occupy during my stay there, a corpsman was taking my pulse and chatting about the Army. A nurse walked in, with a face and figure of such surpassing beauty that I forgot my manners and just stared. I hadn't seen a woman of any kind in eight weeks and she was grade-A fantasy material right out of Playboy magazine, with a sweet voice and manner. 'This must be what Heaven is like,' I thought. The Corpsman looked up annoyed, and said, "Hey, get out of here. This guys's pulse just doubled."

She dimpled a smile and left the room.

"Take a deep breath," the Corpsman said. "Calm down and we'll start over."

"Sorry," I said. "It's just..."

"I know. I'd like to fuck her myself," he confessed. Not everyone in MSC was gay.

"Isn't she an officer?"

"What does that have to do with it? No. She's a civilian. Local girl. And married with kids. Besides, what's being an officer have to do with it? Some of our parties here give Roman orgies a fair run for their money. You have no idea."

"Actually, I do. My father is Chief of Surgery at Brooke. Full Colonel. Threatened to disown me if I went MSC."

The Corpsman already knew who he was, and seemed less than impressed.

" Has he ever been in a war zone?"

"First MASH during the Korean Conflict."

"Then he should remember what it's like. But he's a hardass, I've heard."

"Yeah." I replied, and left it at that. The Corpsman, I later found out, had done two tours in Vietnam and was getting out after eight years of service so he wouldn't have to do a third. He was one of the good guys, brave enough, but he'd gotten his bottle full.

The surgery the next morning went well, and I came out of the anesthetic with the worst belly pain I'd ever had, and a nine-inch scar on my right side. It was state-of-the-art surgery. It was also still 1967. I was only semi-conscious because of the painkiller I'd been given.

As I lay in recovery, and then in my assigned hospital bed, I became aware of an exciting new television program that seemed to play all day long. It was called the Six Day War. It was the latest example of the neverending story of a hot news cycle. It was hard for me to pick a side. I was very pro-Israeli, but also felt great anger when the news about the attack on the U.S. Liberty came out. The clerk at the ASA personnel attachment had called to check on me and, in the course of that conversation, enlightened me.

"'Intelligence ship' means SIGINT, Hamit. NSA, like us. They might be Navy, but they're more like us than the Regular Army. Those are our brothers who got shot up." He sounded quietly furious about the matter and it made him more than just a disembodied voice on the phone. He might be stuck in a boring job at a post in the Midwest, but he was still 'ASA All The Way'. For the first time since I'd enlisted, I felt like I was part of something: a unit with a real job to do.

Reddy Lemons stopped by to examine his handiwork. I was made to walk by the corpsmen on the ward hours after returning to my regular bed. One on each side, the other being the cute, self-professed bisexual African-American girl. I almost passed out from the pain, but they told me it was for my own good, and ignored my curses. Now Doctor Lemons had another concern.

"You need to cough more to clear your lungs," he said, his manner severe. "Get on it, Private, or we'll have to give you a lavage. I know it hurts, but suck it up and be a man." He walked off rapidly to see more patients, and I looked at the Corpsman, who was becoming something of a buddy, and asked, "Lavage?"

"They pour salt water down your throat and make you breathe it. It's like drowning."

Ouch! I didn't want that, so I 'got on it' and started trying to get the mucus and other crap out of my lungs. It really hurt, but drugs were provided. They dulled the pain a bit and made me foggy, but, then, I wasn't going anywhere, and didn't have anything else to do. I couldn't even read, much less write. My manuscripts sat neglected.

A friend of mine from Iowa City, who'd created a new color slick magazine called The Stud-Ent sent me a letter. The title was a bad pun. (Student. Stud-Ent. Get it?) He sent me the first issue, which had a cover illustration of a meter-maid writing a ticket overlaid with the cross hairs of a rifle scope. Meter-maids were generally hated in Iowa City, because the tickets they wrote were a major source of income for the city, which lies right next to the University of Iowa campus. The parking fines were ruinous.

It was funny in a sick kind of way. Another feature was a centerfold like those that had made Playboy magazine so popular. And the topless and generously-endowed blonde it portrayed was a local sorority girl. In 1967 this was shocking. Not a remote fantasy figure, but a girl you saw every day on campus.

With the magazine, came a letter from my friend, asking me to write a humorous article about Basic Training. I wrote back that there is nothing humorous about Basic Training. My first journalistic assignment, and I turned it down.

I was about two weeks into my recovery when the Corpsman told me a particularly lewd and funny joke. I laughed so hard I busted a gut. Literally. My incision, which was healing nicely, ruptured. So it was back to surgery to fix the problem and another two weeks delay. I think the ASA personnel clerk simply threw up his hands at this point and, seeing my 123 score on the radio code test, put me on the list for the Intelligence school at Fort Devens, Massachusetts, and let it go at that.

I came out of the second surgery screaming my head off. The painkiller used, Demerol, alleged to be three times stronger than heroin, simply did not work that time. I was also using language so foul that a civilian nurse from the local community kept yelling at me to: "Watch your language!" It was apparently an impressive display of invective. A few days later, a kindly Nurse Corps Major named Butts dropped by to see me, because she'd promised my father to keep an eye on me. They'd been at First MASH in Korea together during that war. She had a twinkle in her eye as she mentioned the cursing.

"You know, Hamit, I've been in the Army for 22 years, and never heard language like that. You're going to be a great NCO."

"Thank you, ma'am," was all I could say. I couldn't possibly be disciplined for what had happened, because I'd been on drugs, and therefore not responsible for my actions. If I'd had any ASA training at that point, someone from ASA would have scrubbed in to observe and monitor my speech, but there was as yet no need, because I didn't know anything classified.

That civilian nurse was very upset and wouldn't speak to me, calling me a 'potty mouth.'

She was there only because the Army didn't have enough Nurse Corps officers to fill all the slots and a nurse is a nurse. The Army employed almost as many civilian workers as it had soldiers and that's true to this day.

What had happened? I am one of those rare people who builds a quick resistance to opiate painkillers. Because I was given another heavy dose of Demerol, I couldn't be given an alternative painkiller such as morphine for several hours. I lay in agony until they decided that another shot wouldn't kill me. Had I not been 'Colonel Hamit's son' they might have been quicker and taken the risk.

CHAPTER 4

I spent a month in that hospital, and rather than going immediately to Fort Devens was held there another month in my Basic Training Company as an out-patient. They were at something of a loss about what to do with me. I was on 'light duty' again, but couldn't be mixed with the new trainees nor help with the training. I was given a bunk in an empty squad room on the first floor by myself so I wouldn't have to climb stairs. The First Sergeant renewed his campaign to steal me as his clerk, somehow thinking that the choice was mine to make, and I was assigned to help out in the Supply Room. I couldn't type, but I knew the alphabet and could file, and the Supply Sergeant, a genial guy with a blonde buzz cut and black framed Army glasses, made me welcome. I was amused to hear the DI who'd abused me and called me a weakling, now describe me publicly as a model soldier who was so tough he could finish the cycle on crutches.

"If he can do it, you can do it," he shouted.

Their methods hadn't changed much.

A bayonet went missing after a training exercise, and everything stopped until it was found, which took several hours. The year before, a DI had been found on the parade ground at dawn with a bayonet in his belly. Quite dead. The killer was never found. That episode haunted those DIs and they were near panic. My sympathies were with whoever had appropriated it. I had gotten through bayonet training myself by imagining my DI and ADI on the receiving end of the thrusts.

As I healed, the Supply Sergeant and I became friends. He was a career soldier, an E-6, had been to Vietnam, and had some interesting stories to tell.

One day, on his way to his new assignment in Washington, DC, my father dropped by. He was in civilian clothes. I was not very happy to see him. I saluted him and he returned it, with an embarrassed look on his face as if I were eight years old and doing something inappropriate. My mother and sister were in the car nearby.

He had a professional interest in my scar and had already dropped by the hospital to look at my medical records and talk to Doctor Reynolds and Doctor Lemons, whom he described as 'very sharp'. I reluctantly unbuttoned my shirt and let him examine the still-healing angry red scar on my right side. He could find no fault with it. It was, as another doctor said during a later examination, 'elegant work'. He was both relieved and annoyed, and I was annoyed with him. He offered to take me to lunch with the family, and I was graceless enough to say that my duty day wasn't over.

The hapless 1st Lieutenant who was still our Company Commanding Officer and would never be promoted to Captain because of the way he'd mishandled the trainee beatings and the moron I'd been assigned to supervise, came sneaking in to meet him, oily and sucking up like a road-company Uriah Heep.

His negative O.E.R (Officer Efficiency Report) from our kindly Battalion Commander was anything but kind. It became legend on the Post. His career was effectively over, and he had only to serve out his time before being released to a reserve unit near his hometown. I think that Dad saw that everything I had written about the man in my letters home was true and then some, and now wanted to leave before things got really embarrassing.

He and the rest of the family went away. I politely refused yet another invitation to spend some leave time with the family in Washington. My orders for Fort Devens were in, and I was to report ASAP. Five months in the Army and I was already behind.

CHAPTER 5

At Fort Devens, I was assigned to C Company. It was the transition company for the training regiment, which had individual companies lettered down to 'H'. Each company was oversized and had as many as 500 men, all at various schools. 'C' stood for 'casual' and it pulled most of the details for the rest of the regiment. Especially KP.

Col. Lewis Millett

'Kitchen Police' is a long-standing Army tradition. Ordinary soldiers assisted the cooks in preparing and serving the three meals a day required. I had pulled it a couple of times at Fort Leonard Wood. Since I had actually worked in restaurants, I could usually get myself assigned to the Clipper, a big mechanical dishwasher that steam-cleaned racks of plates and glasses which were allowed to air-dry before being put away. Wiping them down with a towel was forbidden. It had a higher risk of contamination and any mass-feeding situation carries the risk of massive food poisoning. I saw this a couple of times in the dorms at college. KP involved constant cleaning. The Clipper was a choice assignment, but one that required a lot of physical effort.

I had to be careful of my incision, even after going off 'light duty' status.

There was a rumor that KP would be taken over by civilians, but that never happened. There was funding, but the funding was diverted to another program: Colonel Lewis Millett's Tactical Training Course.(TTC). He was the Commanding Officer of the school and the training regiment.

He won the Medal of Honor during the Korean War for leading the last bayonet charge in US Army history. He and his men cleared a hillside of hardened Chinese Army Regulars. Before coming to Fort Devens he'd been in Vietnam, where he was part of SOG, the Special Operations Group. This was the behind-the-lines organization that was as much CIA as the Army.

His experience there worried him about the state of readiness of the men under his command if they were captured by the enemy. He knew, first hand, that the Viet Cong and NVA did not acknowledge, much less abide by, the Geneva Convention on Prisoners of War.

Between 1965 and 1973, when it was dismantled, most of the trainees at Fort Devens, whether or not they were bound for Vietnam, went through between three and ten days training at the TTC.

It was in a large patch of woods and included two 'Vietnam Village' sites and a 'POW Camp.' It was staffed by a select group of trainees of Asian heritage, mostly from Hawaii, called the 'Menehune' Platoon (named after a mythical character from Hawaiian folklore much like a Leprechaun, pronounced 'mini-who-nee'). They were assisted by 'Aggressors' (simulated enemy soldiers) drawn from the ranks of C Company. This was a prized assignment. It got you out of KP.

I volunteered for it several times because, again, you were a kid playing war, carrying an unloaded M14 and running up and down the hills and through the woods during the Escape and Evasion problems. Much more fun than being a kitchen drudge.

The Menehunes had their own barracks (which, if you think about it, was the kind of racial segregation the Army had foresworn 20 years before), and special privileges to compensate them for doing double duty. They were all students in various schools as well – except for the female artist who played a Viet Cong interrogator at the P.O.W. camp. She joined ASA to see the world, and found herself stuck at Fort Devens for her entire four year enlistment, making training aids and charts. She did find love, and married one of the Menehunes.

'Vietnam Village'

The first time I was an Aggressor at the TTC, the trainees were not ASA, but ROTC students on a summer drill. They'd had the full ten days and it was a hot Fall day. As they broke out of the Vietnam Village to begin their Escape and Evasion problem, a bunch of them came charging up the hill where I was part of a four man fire team waiting for them. Since we were playing guerillas, we had only our unloaded M-14s and no field gear or canteens. As they panted their way up the hill, we took them prisoner. It was too easy.

I pointed at a line of them. "You, you, you, you, and you. You're captured."

To my amazement, they all dropped their weapons and raised their hands. Suddenly, I had eight prisoners. Thirsty, I demanded a canteen from one of them, and one of my more experienced squad mates cautioned me to make them drink first. Some wiseass on a previous problem had filled his canteen with urine. So each of them got a drink on the house before we took their canteens away. We did body searches, with one squad member holding a rifle leveled at each of the captives while two others patted him down. It was rough treatment, but no less than they deserved, we felt, for giving up so easily. And not a patch on what the Menehunes would do to them at the P.O.W. camp. Those unlucky enough to be captured were tortured.

This is prohibited by the Geneva Convention, of Prisoners of War. The VC and NVA were not signatories, so it was permitted in the name of realistic training, although it was also a violation of both Army Regulations and common sense.

The official manual for the TTC was provided to me by an Army historian. It says:

> *"Once captured, the students are escorted to the POW compound where they are put through a realistic interrogation, complete with 'torture devices' to endeavor to get them to 'talk' (every device in the compound has been tried by each member of the cadre prior to its use on any student)."*

John Wadje, the non Hawaiian Menehune said of the torture; "I couldn't bare to watch it. Much less do it". He subsequently served three tours in Vietnam.

The course was comprehensive and included infantry tactics, land navigation, familiarization with the M60 machine gun and the M79 grenade launcher, and simulated encounters with Vietnamese villagers who might or might not also be VC. It was kind of a military Disneyland.

Whether or not you ended up in the POW camp was luck of the draw. If you were a skilled woodsman, it was easy enough to evade the aggressors who were just other trainees detailed to the TTC for the day. If you were unlucky and fell into the hands of the Menehunes, under the close supervision of Millett himself, then you were exposed to a full menu of tortures that included being placed on an Apache pole, confined in a metal locker that was beaten on, and electric shock from an old field telephone set. The piece de resistance was waterboarding.

Not every guest unit went along with this nonsense. A platoon of Navy Seabees (Construction workers with weapons and Marine Corps training for combat) were sent to get the benefit of Millett's special training. Allegedly, when one of them was placed on the Apache pole, the Chief Petty Officer who was their NCOIC loudly objected. No one was going to abuse his men. He reserved that privilege for himself. He drew the line at torture.

"Stand down, Chief," Millett said.

"No, Sir. I will not!" replied this senior sailor, an E-9. He called for his men to all rally to him, formed them up, and marched them out and back to their temporary quarters, where they packed their gear and left. He preferred charges against Millett but nothing came of that.

Outside units were accepted at the TTC to help justify its existence. That was a mistake.

After a time, word leaked out. An article in the Boston Herald inspired a query from Senator Edward M. Kennedy. The three page letter sent in response made a very strong defense. It cited Millett's impressive military record and decorations. The Medal of Honor, the Distinguished Service Cross, the Silver Star , two Bronze Stars, the Army Commendation Medal and four Purple Hearts. The escape and evasion exercise was six hours of a ten-day course. It was designed to prepare the students psychologically for the war zone.

What it did not do is prepare the Menehunes, some of whom developed psychological problems. PTSD was the least of it. One became so convinced that he would be killed by American troops in Vietnam because of his Asian appearance, that he was assigned to West Germany instead, when he completed his school.

As I said, Millett's previous assignment was SOG in Vietnam. His fears of Viet Cong and PAVN torture were not overblown, and SOG was a primary consumer of ASA product, especially for the missions it ran into Laos, Cambodia, and North Vietnam. One ASA unit was also an airborne special ops detachment.

Millett's defense of the torture was that it was no worse that the physical pain from a gym workout. He disputed accusations of starvation and extreme exhaustion. He emphasized that ASA was an all volunteer unit. No draftees were injured at the TTC. Men had to be trained for overseas duty 'to insure they may live their full life and not die in some stinking rice paddy.' That last is an addition to a prepared briefing in his own handwriting. He was passionate about the need for such training.

Menehune ("Little People") Platoon of Hawaiian Asian-Americans Aggressors at the TTC
The Menehune's played the part of the VC in the John Wayne movie "Green Berets"

Simulated Vietnam Village at the TTC

In view of Army Regulations and CONARC requirements, it is my considered opinion tht I would be derelict in my duty if I recommend discontinuance of the Tactical Training Course. In addition because individuals of ASA are not trained in the normal AIT (Advanced InfantryTraining) subject matter and therefore more extensive and intensive field training to prepare them for combat areas as needed, I cannot concur with the recommendation to stop training. Therefore it is my recommendation that ASA headquarters, Arlington Hall and the Post Commander should be notified that this training establishment should receive official cognizance.

I contacted the lone female member of the Menehunes, Gwen Eloise (Cortez) Kaneshiro. Here is what she wrote me about her TTC experience and what she observed.

The Menehunes were never sworn to secrecy by Millett. We got to know his family and even built a getaway shack on his property in Maine, where they treated us like family. The colonel was nice, but a little bit crazy. I never saw Col. Millett torture anyone, but he often dropped into the camp to make sure we were abiding by the rules. We were not allowed to intentionally physically harm anyone and torture was not allowed to last longer than 20 minutes, however sometimes we were so busy we had to put prisoners in cages to await their turn. None of us were allowed to administer "torture" until we had ourselves endured that same torture. I, too, know what it feels like to be put on top of the "Apache Pole" and wired up to the field telephone for "a call to Hanoi."

Another Menehune veteteran sent me this online several years ago:

After almost 50 years, I don't have much more specific information to offer. I did not go through or see tortures like Kerry describes the buried storage locker or roasting spit. We were pushed around, occasionally tripped or knocked down in the mud/snow and verbally harassed (not much different from basic training), never physically beaten. Tired, cold, wet, and scared – it was a very intimidating experience. And there was the electric shock from field telephones and the poles around which they wrapped your legs before bending you over backwards.

The electric shock was not pleasant, but nowhere close to the pain caused to muscles and joints by the pole. When they were demonstrating how to use the phones on those of us who became interrogators, they carefully explained about not putting the contacts on the body in any way that provided a path through the heart – both contacts on one hand, never one on each hand. Ankle and knee on one leg or finger and elbow on one arm were the most common. Never around the head or genital area. I had damaged knees from sports injuries so the pole was the worse thing for me. After about 15 seconds I explained in loud, clear, and very profane terms what was going to happen to the guy doing the demo if he didn't let me up. He did but certainly might not have been as easy on a trainee. I was probably a lot more sympathetic in my few times as interrogator when using the pole. I never did use the phones. They were usually operated by more senior staff.

Most of the "prisoners" received the harassment treatments while witnessing others being shocked or wrapped around the poles. The intimidation factor of knowing that you might be going there next was sufficient to make even the tough guys very meek. As I understood it, the goal was to make them realize that as bad as it was, what we could do was not even close to what would happen in VC hands, so surrendering other than to avoid certain death was not an option. Certainly, everything I've read about VC and NVA interrogation techniques make that a very true statement.

Millett took full responsibility for this even as he threatened to retire. (Hard to do in the middle of a war. An idle threat.) SOG's soldiers were highly trained and motivated. ASA's not so much, because of the deceptive recruitment that brought in many men who felt, not without cause, that they had been tricked into sacrificing two additional years of their life for military service during an unpopular and badly run war. Attitude problems abounded.

Millett's heart was in the right place. He was a bit of a character with a big mustache and a very ugly dog that he took with him everywhere. Those that worked for him called him kind and courteous. The TTC was improvised theatre from the start, built from scrap metal and wood, weapons borrowed from other units, and vehicles whose next stop was the junk yard. Not all of the actors and audience applauded the play.

The TTC experience, after interrogation, ended in a locked hut with an easy escape route, so the course could be completed by those who were captured.

Discussions of the TTC experience on ASA Facebook groups revealed how varied the experience could be from class to class. One ASA trainee, deciding that he'd had 'enough of this bullshit' picked up a large tree branch and simply walked down the road, promising to attack anyone who tried to stop him. Others, who'd played aggressors on previous problems, knew the course and had the home-field advantage. They were seldom captured.

Those of us who captured the ROTC students were disgusted at how easy it was. These guys would be leading men in combat after they graduated. We lined them up, with their hands clasped behind their necks, and marched them down the other side of the hill, trying to make sure none of them fell and injured themselves, and hindered by the two or three extra rifles apiece we had to carry for them. When we got to the dirt road that led to the POW camp one of them made a break for it. He and a few others scattered and ran away, all going down the road in the opposite direction from the camp. The one non-Asian Menehune, a full-blooded Papago Indian, inevitably nicknamed 'Chief,' ran after the instigator. He was very fast and very powerful, over six feet tall, but his quarry was no slouch in the running department either, and got about a mile and a half down the road before being recaptured. He was made to low-crawl the entire distance back, and the extra half mile to the camp as well. Some of his mates went into the woods and were blundering about there.

After we delivered the rest of the several dozen ROTC students we'd captured to the tender mercies of the Menehunes, we went after them. I sat concealed behind a bush, taking a break, as various ROTC students crashed around the woods, picking mosquitos off my face so as to stay unobserved. When one of them got too close, I simply intoned, "Good Evening," in a deep theatrical voice, and he would go crashing off in a different direction. In the ninth and tenth hour, with little water and no food, it was no longer fun, but I didn't know where I would be going after I finished my school, and Vietnam, despite that notorious, lying recruiting poster, was a one-in-three possibility.

As far as I know, none of the thousands of men who went through Colonel Millett's Tactical Training Course were killed or even seriously injured. Some thought it great training that they might need, and others endured it. By the time I took it the following February, it was a much different experience, which I will narrate below.

It was hard on the Menehunes. They were allegedly sworn to secrecy, and most of them still abide by that. They were also students, but isolated from their peers in that segregated barracks, and working the POW Camp every week in addition to their classes.

A few years ago, I helped one of the Menehunes file a claim with the Veteran's Administration for PTSD. He was never in Vietnam. In fact, he was the one that became so convinced that he would be shot for his Asian appearance in a Vietnam duty post that he was sent to one of our bases in West Germany instead, and finished his enlistment there.

His memories and guilt created so much PTSD that one of his children contacted me for help. There was little on the official record of the TTC or the Menehunes, and it took me quite awhile to find documentation to support his claim. It was granted. The INSCOM (successor organization to ASA) Historical Department has little documentation.

One of the facts I was able to point out for the guy I helped with his VA claim was that some of the Menehunes became, at the behest of the Army, movie actors.

"The Green Berets" a movie about the Special Forces, was a laughable farce with the infamous final scene where the sun appears to set in the East. No Vietnamese appear in that movie, so the Menehunes were assigned to play the various Viet Cong seen in some shots, and were sent TDY at Army expense to Fort Benning, where the film was shot. Because of union rules, they were also paid by BatJac Productions, John Wayne's film production company, and housed off base at a motel. They also played other Vietnamese roles and some even had minor speaking roles. The US Army supported and influenced the film to an extraordinary degree. It was their counter-propaganda to the anti-war films made in Hollywood at that time.

What role Colonel Millett might have played in this remains unrevealed, but he was accused of using ASA enlisted men to do work on his farm in Maine by some who were at the TTC. Winning the Medal of Honor will buy you a lot of slack. Even Generals salute that medal.

The TTC and its Vietnam Village and P.O.W. camp were bulldozed for a new military housing development in 1973, once we pulled out of Vietnam in 1972, having militarily defeated the NVA forces and eviscerated the Viet Cong.

Colonel Millett retired, and became a Deputy Sheriff. He was too colorful to become a Brigadier General in an Army grown more corporate over the years. Advanced degrees counted more than battles won. He was allegedly also very disheartened by our reluctance to keep fighting that war, or to even support the Vietnamese Army we stood up in our place. In this, he was far from alone. I've never met a Vietnam veteran who does not feel the same way. That war was not lost on the battlefield but in the streets, or, rather, on the college campuses, of America.

CHAPTER 6

Radio code school was a disaster for me. Unknown to me, and everyone else at the time, was the fact that I am dyslexic. The term itself was not in general use in 1967, or my two courses in Remedial Reading in the Third and Eighth grades that might have given a hint. It's basically a hearing problem, and to learn radio code and copy it at high speed, you have to hear particularly well. ASA knew this, going back to World War II.

The minimum speed to graduate from that school was making a copy of a message of 18 five letter groups per minute with 85% accuracy. You got extra points and possibly a promotion if you could do 25. You had to type these groups on a special typewriter called a "mill" which had only capital letters. Points were deducted for every error, so that you might actually have a negative score far below zero.

I was selected for the new computer-based instruction which had a computer keyboard and a monitor in a self-contained console with a headset for listening to the code. The instructors were expert intercept operators who'd worked the MOS for years, and had a bad case of what the US Air Force calls 'silk scarf syndrome', after the pilots who resist cockpit instrument improvements on the theory that making it easier somehow lowers standards. The Army has its own version of this, and the FBI back then suffered for years from the lack of computer skills by many of its male agents because 'real men don't type'. You get the idea.

As I struggled to hear and hit the right key for each letter, I also had this flaming egotist of an E-6 who would parade up and down the aisles of the classroom between these large, heavy training machines that had set the government back a nice six-figure sum each, and suddenly try to break a student's concentration by taking his drill instructor's swagger stick and slamming it on the top of the monitor where your typing was displayed. This machine was supposed to be teaching an accelerated course and produce graduates in record time. It was supposed to eliminate the instructors entirely. Perhaps he resented that. His actions interfered with, rather than enhanced, the high-speed instruction.

In the barracks, the students complained about similar tricks designed to throw you off your copy, played in the regular classes in I Division, including setting fire to the copy paper in your mill. Such hijinks were impossible with these new advanced consoles, which had no paper and scored your product automatically at the end of the exercise.

My scores were consistently the lowest in the class, and once hit a negative figure so high that it was placed on the classroom bulletin board. This act of shaming me, and the counseling I got about "you can do anything you think you can do" and positive attitude (the man had read the works of Norman Vincent Peale, a famous motivational speaker of that era), just made me more desperate and depressed. And the first time he slapped that swagger stick down, it sounded just like a pistol shot, and suddenly I was momentarily back in Iowa City, diving for cover. No one had heard of PTSD then, either.

Any kind of mental disability in the Army is considered shameful, and while the official attitude is more enlightened, the cultural one hasn't changed all that much. I went on sick call to get some meds to calm me down, as the good Sergeant suggested, and wound up talking with an E-6 'psychological technician.' This guy had a lot of time on his hands because of the stigma associated with his field, so he signed me up for counseling three times a week. This was treatment I could not refuse. Maybe talking would help. I outlined the entire problem to him with Sergeant Swaggerstick. He listened carefully and murmured comforting words, but said meds wouldn't help. I couldn't hear the code on the machines, and that simple explanation simply was not good enough. Instead, I was viewed as a bad soldier and someone trying to get thrown out of the Army.

After the third session with this guy, who told me that he intended to get his doctorate and become a real Psychologist someday, Sergeant Swaggerstick kept me after class and repeated to me, almost word for word, everything I had told him, and again preached from the works of Norman Vincent Peale. On my next appointment I told the Psych Sergeant that he was an unethical sonofabitch who had betrayed my confidence, and I would take a court martial before I would waste time with him again. I threatened to go to the Inspector General if the appointments continued. He looked embarrassed, and couldn't meet my eye. That was the last we saw of each other.

There came a day when Sergeant Swaggerstick slammed his toy on my console one time too many, and I tried to throw it at him. It weighed 3,500 lbs, and I had it part way off the floor before they grabbed me and quickly took me out into the hallway. When Swaggerstick came out, with this really scared expression on his face, I decided that my original intention to murder him on the spot was not the course of wisdom, and I just stood there shaking. After my breathing and heart rate returned to something close to normal, I was told to return to my company Orderly Room and report to my First Sergeant.

This was Jordan Brown, an easy-going professional who didn't seem surprised to see me when I showed up, shamefaced and fearful for what I had done. My console was trashed because it had been dropped six inches, but Jordan Brown blamed the instructors for pushing me too far. Episodes like this were not unknown in the regular I Division classes. Mills flew through windows about once a week in those classrooms, and those were guys who got frustrated at far slower speeds. No punishment was ever administered for losing it in a code class. You were there because you already had the very extensive and expensive FBI security clearance for Top Secret/ Crypto, and if you washed out of a school, there were plenty of OJT jobs available within ASA. Even the cooks needed that clearance. Anyone could learn to cook, and I already knew how, having worked at several restaurants in Iowa City as such.

"Until I get you in another class, you'll work here as my runner," the First Sergeant said. "Be here at 0800 tomorrow and look STRAC." That meant in a fresh, starched fatigue uniform and a good attitude. I took extra care the next morning with my appearance.

Brown was disappointed to learn that I couldn't type, and sent me off with messages for various people during the day, but otherwise it was easy duty. I even got to read, very carefully, his copy of the New York Times, which he had delivered every day and paid for himself. It had to stay pristine because he saved for his evening hours at his quarters.

But he was kind, and he was trying to see if ASA should keep me. So for two weeks I made myself as useful as I could, and he concluded that, rather than me failing, the system had failed me, by placing me in an advanced class I had no hope of passing, based solely on a test score that regularly failed to select men who could copy code.

He said as much to the Company Commander, Captain Crybalski, whose mantra was "Never assume. The word 'assume' makes an ass of you and me." Crybalski said this so often it became a G Company slogan. But he was Commander in name only. Jordan Brown ran the show, and everybody knew it. When there was a showdown later that year between them over an issue never revealed to the rest of us, it was Crybalski who moved on, exiled to the TTC, and Brown who stayed. ASA was run by its NCOs.

My barracks held code-school students, and the barracks next to it held those for the mystical dark arts of Electronic Warfare. This was the toughest school, intellectually, the ASA had. It was equivalent to Navy SEAL training for the mind, requiring heavy doses of Ph.D. level math and physics. You had to be really smart to get into ASA. You had to be a super-genius to become a 98J Electronic Warfare Specialist. I got along with those guys very well because of a mutual interest in science fiction and fantasy. They were fun to talk with, and sometimes I would adventure with them on weekends to Boston and New York.

The Electronic Warfare mob had its own barracks floor, and a mythical mascot named Oscar. Oscar was a duck, eight feet tall, wearing the ASA patch on a t-shirt, with size 32EEE tennis shoes over his feet. There was a large cartoon representing him painted on the outside of their barracks, which was next to the one I was in. In class, when the lecture got too boring, they would begin to 'talk to Oscar' by making quacking noises like a duck. This drove the instructors crazy, which just made them do it more. It was a not-so-quiet rebellion. They were under a lot of pressure. It made them a bit wild and crazy. They needed to blow off steam just to stay sane.

The stuff they were dealing with was so secret it had its own codeword, and some of the rest of us got a hint in a bull session one night when one of them dropped the phrase 'laser radar.' This was 1967, and such things were the stuff of science fiction, except, apparently, not. We can do that? I thought. The reaction from the 98J class men was like a fart in church. A major security breach had just occurred, but no one wanted this guy to suffer the consequences of his loose tongue, which could include jail time. So, after a few seconds, we all continued as if the words had never been uttered. Nor did anyone ever mention again the technology now called 'Lidar'.

They were fun to go to Boston and New York with. One of them would go to the Public Gardens and quack at the ducks, and the ducks, to our amazement, would answer him. What Top Secret program was that from? There was a long street leading from the Public Gardens and the Commons to the river, and it was a place to try and meet girls, except the Army haircuts made us all marked men. The war in Vietnam was having an impact on our social life. There was also the notorious Combat Zone next to the bus station, and that had some bars and other amusements, but Army pay didn't go far there.

I remember sitting in a booth in a deli one afternoon listening to folk singer Phil Ochs strum his guitar and put on an impromptu free concert. I was too shy to thank him.

One weekend I went down to New York City to meet up with a couple of girls I had known in Iowa City. They were friends of my former fiancee. One of them was living there, and they both dressed up to look glamourous. Apparently they overdid it. I took them to a restaurant where they were taken for hookers, and I for their pimp. That rather spoiled the evening. The truth was that one was married by then, and the other about to come out as a Lesbian. And they were not pleased I was in the Army, in Military Intelligence. They tried to persuade me to desert and go to Canada. I declined, and pointed out that there were other places to serve besides Vietnam. It was the last time I saw either of them, which was sad, because I needed friends.

Weekends at Fort Devens were free time, unless you caught Charge of Quarters or a fire detail. The barracks were World War II vintage fire traps, so coal had to be shoveled to heat them, and fire watches had to be set to keep them from burning down in the middle of the night. Long lines of bunks in two rows with steel lockers to one side and a wooden footlocker at the end of the bed were standard Army arrangements, as were the GI parties, weekly inspections, and morning formations. While others might complain about the 'Regular Army BS,' and our resident Staff Sergeant had his own room on the second floor, and assured us that most ASA posts didn't have time for these quaint customs, I was happy to be there. My boots shone brightly. So did my smile.

WW2 Barracks

CHAPTER 7

I was less happy to be sent back to code school. But this was the baby-steps version, with rote memorization of Morse Code, shouted at the top of our lungs to drive it deep in our minds, and then typing. The first step was to meet the Army typing standard of 25 words per minute without error. Touch typing. Fortunately, and perhaps deliberately, we were tested on the same piece of text until we passed, and I finally did so by memorizing the damn thing! Even a blind hog can find a truffle.

Then we started on code at three and a half words per minute, and to my delighted surprise, I not only passed but got an almost perfect score. Perhaps I could learn to do this. But as the rate went up, my scores went progressively down. No one tried to 'motivate' me. They didn't want another breakdown. They just put me in the dumbbell class that worked an extra three hours at night.

Every morning we would form up outside the long row of G Company Barracks, listen to a few announcements read by the student who was given the task of being the Company Formation Leader, and march to the classrooms housed in identical barracks about a mile and a half away. There was no precision, and it was mostly route-march. Getting there in an orderly fashion was the point, not performing precision drills. It was early, we'd had breakfast and had nothing to look forward to but hours if trying to copy code. No one sang. We shuffled along, thinking about how to pass that damnable course.

One morning a crisply uniformed Second Lieutenant saw our ragged formation, jumped from his car and shouted. "Y'all are marching like a bunch of damned niggers!" He then took over the formation and double-timed us down the road for about two miles, and made us march back to a cadence, showing us how it was supposed to be done. Regular Army and new to the post – he looked like a recruiting poster. Very handsome and wearing tailored fatigues. Not West Point, he loudly informed us, but a graduate of the Citadel, "The best damn military school in the world."

No one responded to this because we hadn't done anything like this since Basic, and were all out of breath. We didn't do morning PT.

We were all late for class, sweating and exhausted, and the instructors, all enlisted men, could say little to him because he was an officer. One who used the 'N-word' with abandon. Suddenly the Civil Rights struggle, with all its ugliness, was right in front of us. It was not something we talked about, because politics and religion were not discussed in the barracks. Everyone had an opinion, of course, but expressing them led to arguments and even fist fights.

The Army was eager to reform itself on racial matters. It was policy. But many soldiers and officers from the South, where Jim Crow laws were still in force, resisted. A couple of 'good ol' boy' senior sergeants at Fort Devens had been denied reenlistment after 17 years of service because they couldn't stop themselves from using that word – that meant they wouldn't get their 20-year pensions.

ASA had never had many African-Americans in its ranks, but the ones they did have were very smart, and always seemed to have found positions of power and influence. When the Army finally got serious about racial equality, they led the way.

This Lieutenant was another avowed racist who hadn't gotten the memo. His hasty action came back to bite him. He had violated the chain of command by seizing control of the formation from the student who led it, and had, for that 30 minutes, the temporary rank of Staff Sergeant. He had taken several classes of students on a morning run that exhausted them and made them unfit for instruction for the rest of the day. And he used language and acted in a way that would get him dismissed from his own Intelligence Officer Basic Course and sent directly to Vietnam. The instructors at I Division sent him a lovely parting gift. A box of Tampax.

There were others who had to be re-educated. The great struggle for Civil Rights was not something much discussed among us enlisted trainees. Most of us were for the changes that had to be made. The exceptions were the guys coming from southern states where the struggle blossomed into open warfare.

I had African-Americans on my stage crew at Tamalpais High School in Mill Valley, and some of them were friends. Tam High was considered the really tough high school in Marin County – because we had Black students. They came from Marin City, which was then the local ghetto, but while some of them seethed with resentment, others welcomed white students to the Struggle, knowing that they couldn't do it alone.

I had to be very careful. My father was a serving officer in the US Army, and my actions would come back on him. And then there was my mother, a great Southern lady who couldn't shake free of the prejudices she'd learned as a child. I'd attended no demonstrations. I might have helped plan a few, but talk is cheap, and I took no leadership role.

Some of us went to Boston one weekend. My buddy Ray was from Alabama, and newly arrived from Basic Training at Fort Leonard Wood, as smart as any of us but also the prisoner of his upbringing. Too smart to betray his true feelings, even after I told him that I sympathized with the 'outside agitators', until that trip to Boston – when he suddenly turned bright red, enraged by the sight of four people walking down the street. Specifically, two black men and two women who might have been white. He was ready to commit violence, because in Alabama this would be a serious matter, one worth killing over.

I took it upon myself to bring him down.

"Ray," I said, "I see that you are upset. So listen to me carefully. Those people are just walking down the street, and you can't assume that there is anything else going on. They don't know you, and you don't know them, and this is Massachusetts, not Alabama, so get yourself under control. No one is trying to offend you, and if you start anything, you will risk everything. Even jail and being thrown out of the Army. Do you understand?"

He stared after the two couples for a long time, and then looked down at the ground. "I guess you're right," he said softly. We said no more about it.

It was about this time that TIME magazine published the cover with the wedding photo of Guy Gibson Smith and his bride Peggy Rusk, the daughter of Secretary of State Dean Rusk. He was Black and she was not just White but a flower of Southern womanhood. Ray's worst fear personified.

Guy and I were classmates for two years at Georgetown Day in Washington, DC, my first African-American friend, when the schools were still largely segregated because Brown v. Board of Education of Topeka was still a very new decision. Guy was an Army brat like me, and would become a helicopter pilot during the Vietnam War. We had a fist-fight once, but generally had gotten along, and he'd been our class president both years. I mentioned this to one or two people only to find the notion scoffed at. How would I, a mere private, know someone so famous? Of course, fame is fleeting, and this brief exposure was as famous as Guy and his wife ever got.

CHAPTER 8

I fell further and further behind in code school. So much so that I became, for a few weeks, the G Company formation leader, marching my fellow students to class every morning, and since I set the pace, I could finally march in step. On any day, it was more a battalion than a company, with about 500 men and a sub leader for every barracks to relay my commands. I tried to make it a little more military. I found that calling a cadence made it easier for everyone, and actually improved, not just the look of the company on the march, but also morale. Someone called me a 'lifer'. I just shrugged. Maybe I was, or would be. I had not yet made that decision. If I could only get through this damned school. I also marched them back at night, holding them in formation until dismissed, rather than just letting them scatter toward the Mess Hall.

Fort Devens in Winter

There was one senior NCO who declined any leadership role, and every night, I would begin my dismissals with 'All E-5s over 20' (more than 20 years service). Shamefaced, he would move off on his own before I dismissed the rest. He had the red nose and downcast look of a career alcoholic, and was highly skilled, I was told. He might have been an instructor rather than a student. Unmarried, except to the Army, he had been busted from E-7 for some drunken escapade, and was being retrained. It was a sad sight, and he probably should have taken retirement, except that being in the Army kept him somewhat sober. It was thought that if he retired, he would drink himself to death within a year. Someone, somewhere, liked him and protected him.

The snows came, and I was tasked with coordinating the snow-removal details one night when I was CQ. I was in and out of the heated barracks several times, and three days later developed a high fever. I had pneumonia, and would be in the Fort Devens hospital for a month, drinking ice water to keep my temperature below 103 degrees.

When I came back to I Division I had to start over and repeated the pattern of early successes followed by gradual failures, stalling out at the 12 groups per minute level, well short of the standard needed to pass the course.

My father wrote: *"Even before you were born I was told by a girl who taught code to would-be radio operators of World War II that there are people who can take code and people who cannot take code. So that bit of "news" was no great revelation to me. And I also know that I.Q. has little to do with it. I am assuming that they decided that you are one of the people who cannot take code, and the Army has finally realized this, having tried to teach you by methods tried and true. I do not question their decision or what you have to say on the subject, but I know your proclivities for scuttling anything that gets down to the nitty-gritties of a little sweat and hard work."*

In this same letter he complained about my lack of correspondence to him, and then informed me that he was putting in for his retirement the next day. He was on the list for Brigadier General. An FBI Agent had sought me out at Fort Devens about his political opinions because he was being given a new, higher security clearance (but not one as high as the one I already had).

He felt that, after 26 years and three wars, he had done enough, and he was being offered a big job at a drug company, one rung below Vice President. He was quitting the Army because of a few minor incidents with his transfer and working conditions. This was where he was in touch with his 'inner child'. It was not a rational end to the career he built. The promotion would have him take over all the US Army medical units in Vietnam for a year. On the ground.

The Army's mistake was sending him as an Inspector General to Vietnam that November, and giving him a good look at what he was in for if he took that job. If they had offered him Europe, it might have been different, but he suspected there was a clique who kept those choice assignments for themselves. He may have been right.

Col. Harold F. Hamit, MD, between stops on his inspection tour in Vietnam, Nov. 1967

And he was right about my record of failures. In high school I had failed several language courses and science courses he insisted I take. He wanted me to follow in his footsteps, and considered my theatrical work a passing fancy, and never credited the hard work and heavy responsibilities that went with being a stage manager. He didn't even understand that I was running the show, and could not hang out with the family before, during or after the performance. This is petty, of course, but it was part of the same pattern of willful disregard on his part that formed our relationship. He set me up for failure rather than let me find my own way.

I volunteered for a war he declined to be part of. Not for any political reason; like every other serving officer of that era, he avoided politics, and any comments he made in his letters were simply further exercises in sarcasm that hit both sides of the political spectrum. He had a keen wit and a jaundiced eye.

As for my own failed career, sitting on a position copying code was not what most of us had signed up for, so most troubled students were trying to fake their way out of an onerous and boring task. It was different once you got to the field, I was later told, because it meant something, and was no longer this boring exercise. It's hard to see how, since the code received was encrypted and had to be broken and analyzed by others. But some people, especially those between 18 and 25 years old, took to it like it was the easiest thing in the world. I had played with Ham Radio when I was a kid, and really did want to learn this stuff, but it was no use. I couldn't hear it fast enough to pass the course. No amount of additional training was going to fix the problem, either. It was not a matter of application, but innate ability, of which I had not enough. Nor was I offered any alternative training such as Traffic Analysis or even the clerk school. They were fixated on getting me up to speed.

Finally the Warrant Officer in charge of I Division, who'd come up through the ranks and could copy fifty groups a minute himself, called me in and said, very kindly, "Hamit, this is ridiculous. You've been here three days longer than I have."

I saw my opportunity to escape, and said, "Sir, I don't think I can do more than I have. I respectfully ask for a transfer to Vietnam."

He smiled, nodded, and said he'd see what he could do.

I was relieved of the school and went back to being Jordan Brown's orderly. It was a very pleasant, low-stress assignment, but a few weeks later my hopes that I could ride out my time in ASA as a company-level clerk there were dashed. Volunteering for Vietnam was perhaps an act of desperation, although part of me felt I needed to see that war, and find out what was really going on.

CHAPTER 9

I was sent back to Company C to await orders, and pull what seemed like an endless chain of KP duty, day after day. This was the fate of all casuals, and some of them, against Army Regulations and all common sense, pulled it for far more than the 30 days a year that was supposed to be the maximum. They were long, hard, sixteen hour days that left barely time for a shower and sleep, and the winter conditions at Con 4, the big mess hall that fed the entire regiment, were brutal, as were the cooks.

Imagine joining the most elite, secretive military intelligence outfit in the Army, and winding up as a cook! Someone had to do it, but to be in ASA, you had to have that minimum GT score of 115, and if you were in that lower range, then you got the jobs that any soldier could do, while the interesting work went to the guys who'd scored 30 points or more higher. It was arbitrary and unfair, because the GT was a poor predictor of how well you could actually learn a job, but it was also the Army, with personnel policies administered on an industrial scale. The cooks were bitter to a man, but no one really blamed them, even as they took it out on the rest of us on KP. The food, let it be noted, was usually the best to be had in the Army. There had been a couple of incidents of mass food poisoning, so they were fanatical about cleaning. Having the entire regiment on sick call with vomiting and diarrhea was likely to send the lot of them to Vietnam.

The two best jobs were the Clipper, or dishwasher, and 'outside man' where you handled the trash and wet garbage and received the food supplies. Both were steam-heated. The worst jobs were 'pots and pans' and the grease pit, which had to be cleaned daily and was just nasty.

There were smaller mess halls in the system, too, but I seldom was assigned to one of those. My last time was when I was newly assigned to the regiment.

I do recall that one of the cooks the previous summer was furtively trying to dry some banana peels on the concrete rear loading bay. He had heard that dried banana peels contained an hallucinogen and wanted to get high. Unlike marijuana, he couldn't be arrested for having dried banana peels. So he told me. I didn't bother to mention my own brief experience as a Narc, because the whole proposition was a silly urban myth. There was no high to be had, so he was right, he couldn't get busted for banana peels.

KP was mildly punishing, but as long as I wasn't driving myself crazy trying to listen to code, I was content. Others grumbled and complained, and one of my fellow victims in the C company barracks took it too much to heart. His school hadn't started, and he'd done 65 days of KP straight before getting a three-day pass and going home to Queens, where his mother had apparently given him a very rough time for his decision to enlist. He came back even more depressed, but I was too tired to take much notice. He was a chronic whiner. 'Woe is me' was his mantra.

The next morning at reveille, as we all started to get dressed for another day of KP Hell, he lay in his bunk unmoving. Someone tried to shake him awake.

"Hey, I can't wake this guy," that soldier said, a little alarmed.

I went over and took a look. His breathing was very shallow. I peeled back an eyelid and looked at his pupils, which were immense black puddles.

I'd seen this in Iowa City. "He's OD'ed," I said. "You two, get him up and get him walking. You, toss his locker and see what he took, and someone get the First Sergeant!"

They scrambled to do as I said.

"Found something," one of them said, and handed me an empty bottle which had contained a hundred Librium tablets, with his name on it. He'd seen a civilian doctor while on pass, gotten a prescription, and had tried to kill himself. I knew that making him walk would keep his heart pumping. If that didn't work, then one of us would have to use the CPR skills we'd been taught in Basic.

The First Sergeant and his company clerk came charging up the stairs.

"What the Hell is this?" he roared.

"Suicide attempt," I replied.

The First Sergeant uttered a string of industrial-strength curses around the words, "Not again!" and, "Call an ambulance!" His clerk ran off to comply.

He saw what I was doing, helping the guy who'd OD'ed walk, and said, gruffly, "You go with him in the ambulance. The doctor will want to know what happened."

I handed him the empty pill bottle. He looked at it and handed it back.

"Not enough. He'll probably live. But keep doing what you're doing."

"What did you mean, 'not again'?"

He leaned in toward me and murmured. "Third one this year. But you didn't hear that from me."

He was the guy who signed the assignment rosters, so I could see that he felt responsible.

The ambulance came, the victim was put on a stretcher, an IV inserted in his arm, and both the First Sergeant and I went in the ambulance with him to the ER, where the doctor on call seemed to be hung over and more than a little angry. He had to pound on the victim's chest to restart his heart. When I showed him the empty Librium bottle, and when he saw the date on it, he tried to match the First Sergeant's curses and didn't fall far short.

He beat on the victim's chest some more, cursing him for a fool, saying he should have seen a shrink first. I had my own opinion on that, which I kept to myself.

Finally, he waved the First Sergeant and me away. "We're admitting him," he said. "You can return to duty."

The company clerk was waiting with the First Sergeant's jeep.

The First Sergeant was very quiet for a few moments, thinking hard. "You were head's up there, Private. Good job. You probably saved his life."

I nodded.

"You breathe a word of this to anyone, and I will keep you on KP every day until your enlistment expires."

I nodded again.

"When you get back, pack up his gear in his duffle, and bring it to the Orderly Room. It all has to be inventoried and secured."

"Yes, First Sergeant. What do I tell the guys about him? They'll want to know."

"That's he's in the hospital but won't be receiving visitors. I'm going to get him discharged."

I must have looked a bit surprised, because then he said, "Honorable. Inability to adjust to service life."

"Yeah. Endless KP will do that."

"Don't be a smart-ass, Hamit. Most guys can handle it. You don't have problems with it, do you?"

"It's not what I signed up for."

"Yeah, and you've fucked up here with the school and all, but it will get better. You're Army born and bred. You were made for this."

"If you say so, Sarge." I had my doubts. It hadn't gone all that well, and I'd been in almost a year by then.

Packing up that guy's gear didn't tell me much about him, except that he was reading a book by Norman Vincent Peale. There was a flyleaf inscription from his mother.

CHAPTER 10

A few days later the Tet Offensive began in Vietnam. The American Army was caught unawares, because General Westmoreland failed to understand the nature of the enemy, and ignored multiple intelligence sources that warned of the attacks. Some of those sources were ours – Signals Intelligence that revealed enemy units in the cities as well as the countryside.

Westy refused to take these 'part-timers' seriously. . He longed for the kind of combat that had made his reputation during World War II. He and some of his staff held that our superior technology and overwhelming resources would wear down the VC.

Even a green Private from Iowa knew it wasn't that kind of party.

American forces and their allies used massive search and destroy operations to deprive the VC of resources. It was never enough. Supplies flowed down the Ho Chi Minh trail, even as far as the Mekong Delta, and the enemy went from using bicycles to new trucks provided by the Soviets and Chinese. Cambodia held huge supply depots for the VC and PAVN forces. Their map included all of the old Indochina. Ours included only South Vietnam. We were constrained by International Law. They flipped it the bird.

While Westmoreland longed, as I said, for the kind of main-force battle that had made his reputation during World War Two, General Giap was playing a very long game. He would pick the times and places of battle. Giap once said to an American correspondent, "You will kill ten and I will kill one, but in the end it is you who will tire of it." In effect, he turned Westmoreland's war of attrition against him, forcing him to expend resources as he sought to close with Giap's divisions.

When PAVN and the VC did give battle, they overwhelmed the USA and its allies. South Korea contributed fighting units, as did Australia, but it was the USA that kept increasing its footprint in South Vietnam. The Marine base at Khe Sanh became a real tar baby for Westmoreland. He became a little desperate and, in my opinion, a lot stupid. Operation Fracture Jaw was his quick and easy way to defeat Giap. He tried to go Nuclear.

He feared a defeat like the one the French experienced at Dien Bien Phu. Desperately, he requested a tactical nuclear weapon nicknamed 'Atomic Annie'. The M65 was a 280mm cannon with a 20-mile range developed during the 1950s. It could deliver a nuclear munition with a 15 kiloton airburst. That's two-thirds the size of the atom bomb dropped on Nagasaki.

Lewis Sorley worked for Westmoreland during the war. He was one of his headquarters staff officers. Later he wrote a book: ***Westmoreland: The General Who Lost Vietnam.*** This book argues that Westmoreland occupied a crucial command in the critical years of the American war in Vietnam that was, in fact, beyond his abilities or understanding. In part, it offers an indictment of the postwar culture of the US Army, showing that Westmoreland's energy, effective selfpromotion, and capacity to 'work the system' led to a command in Vietnam that needed more than unbounded ambition and energy to create success.

Sorley sees one of Westmoreland's basic failings as his avoiding attendance at any of the US Army's major military education institutions; he was, as a result, an uneducated soldier: 'doing it by the book, even though he hadn't read the book or studied at any of the Army's great schools'.

If he had, he might have learned the hard truth about the atomic cannon, or something about ASA and SIGINT.

As it happened, my father was involved in a minor way in the Nevada tests of this monster. He was working for the Army Surgeon General as a research contract officer. He was sent out several times to implant sensors in live pigs that were sacrificed to measure probable human response to the explosion. There were buildings constructed, as well, and some unlucky live troops several miles away. The whole thing was Top Secret codeword. Dad never said much about the results, except that he hoped the weapon would never, ever be used. Nuclear weapons do not discriminate. It would damage us as much as any enemy formations, and poison everything around, with radiation lasting thousands of years. After he died, I found some documents he'd kept that confirmed this. These were the last above-ground nuclear weapon tests we ever did.

I did some research. The atomic cannon was lightly publicized. Six were sent to West Germany and installed near the Fulda Gap to deter a Soviet invasion. One was sent to South Korea to deter an invasion from the North. Westmoreland asked that one be sent to South Vietnam, but not as a bluff. He was actually going to use it north of Khe Sanh!

LBJ found out, and shut the whole thing down, horrified at the probable consequences that ranged from loss of human life, including thousands of American personnel, to China formally entering the war. This was a deep dark secret until 2018.

Operation Fracture Jaw almost happened in January, 1968, shortly before Giap's Tet Offensive the following month. Tet was a shock to the American body politic. The easy victories Westmoreland had promised were not delivered, and Giap's forces came close to carrying the day. It cost him most of the Viet Cong forces and infrastructure, but the comments of Walter Cronkite and other pundits gave him victory where it counted: public opinion. The American public bought into the lie, and that was the beginning of the end.

Westmoreland thought that brute force, in the form of airstrikes and heavy artillery barrages would compel Giap to surrender, or at least back off. It just made him and his troops try harder. Giap had highly motivated soldiers who spent years in the fight, and were going nowhere. Ours were there for a year or less. It was a bloody expensive way to run a war. All those airstrikes cost enormous amounts of money. The ones that the 156th initiated usually were serviced by six B-52 bombers based in Guam. They flew in groups of three, in a delta formation for maximum effect, and were modified to carry 60,000 pounds of 500 lb. and 750 lb. iron bombs. The effect was devastating where they struck, but triple canopy jungle can take a lot of punishment.

Security leaks from our South Vietnamese allies often gave the VC and PAVN enough warning to get out of the way. Giap's own SIGINT establishment was as big as ours, and good at spoofing us.

The disaster of the Tet Offensive came, but not because Westy was not warned. He was, repeatedly.

SIGINT analysts pinpointed the movements and intentions of Giap's forces, and the movement of weapons and materials into South Vietnam's cities. Westmoreland and his staff, focused on his attrition strategy and desire for big battles, ignored the indicators. Only one general officer paid heed.

My ultimate boss, MG Charles Denholm, Chief of ASA, had a field station near Saigon with a thousand men, a battalion of ARDF aircraft, and direct support units attached to every American division. He was no lightweight. Two classes behind Westmoreland at West Point, he'd earned even more combat medals and decorations during World War Two, before joining ASA during the 1950s.

His problem was that his ultimate bosses were at NSA, and they wouldn't let ASA share 'sources and methods'. A lot of product lost all tactical value because it had to be sent to Fort Meade to be analyzed before being released. Everything was compartmented and 'need to know'. ASA officers briefing commanders on the ground shared results only. They did not disclose sources and methods. This led to a lack of trust.

Complicating that was the fact that some ASA units (or, as they were called under cover, 'Radio Research') did not listen to enemy transmissions but our own. 'COMSEC' units were supposed to keep our own radio transmissions up to the mark by making sure proper radio procedures were used. They seldom were. Sloppy transmissions, made up codes easily understood by the enemy, and other unforced errors cost thousands of American lives. Yet, when confronted with their crimes, these commanders reacted like naughty children, and denied the whole thing. Their junior officers and NCOs, rather than mend their ways, called us'buddyfuckers'. They were the ones punished with Article 15s and courts martial. Their loose talking commanders seldom were.

ASA dealt in magic beyond the ken of mere mortals. Seldom did we get the credit we deserved from our customers. That was okay. We were secret, silent warriors. We liked it that way.

Westmoreland did not. He tasked his J-2, Army General McChristian, to short circuit the process whereby NSA saw our product first, analyzed it, and then shared what it liked. Raw intelligence is dangerous and can be misleading when faced with an enemy skilled in deception. McChristian used his rank to bully a group of ASA analysts into giving him first look and letting him decide what was sent to NSA at Hawaii and Fort Meade, Maryland. He forgot that those ASA troops only wore Army green. Their true alligance was to Lt. General Marshall Carter, the Director of NSA and ASA commander MG Charles Denholm.

A senior NCO in that analyst unit went to a public telephone and called Carter direct to let him know what was going on. Carter flew from Fort Meade to Hawaii where he had McChristian fly from Saigon to meet him. With another Army General looking on, he set McChristian straight. The balance of power was restored.

When Westmoreland was promoted to Army Chief of Staff, and finally briefed on ASA's full range of capabilities at Arlington Hall Station outside of Washington, there was notable friction between him and Denholm.

By the time I arrived for my one year tour, there were more than 500,000 troops there, plus contractors ranging from CIA paramilitaries recruited locally to Red Cross 'donut dollies'. It was quite a party, but, with a few exceptions, the average American soldier suffered 'one and done' tours to accommodate all those draftees. The one year tour was a contributing factor in our defeat. No one ever learned their job properly, or, when they did, were soon on their way home.

It was the most pampered army in history. The PX carried a variety of luxury goods at bargain prices. The USO provided entertainers and reading material for free. There were Hollywood motion pictures and television shows distributed on 16mm reels that could be shown on the same Bell & Howell projectors used for training films. (More on that below.)

It was much more than our fathers and uncles had during World War Two and Korea. Despite that, few of us were happy to be there. Morale was low. Stateside soldiers thought that assignment to Vietnam was a fate to be avoided. It was used as a disciplinary threat.

A levy came down for any and all available personnel to be sent to Vietnam – I finally got off KP. But first I had to do the

Tactical Training Course for jungle training – in February, in Massachusetts, in the snow and sub-zero temperatures.

I was issued a heavy woolen winter uniform and hard black rubber 'mickey mouse' boots, more suitable for Korea, and told to wear three pairs of woolen socks to avoid frostbite. We would be camping in tents, too, to heighten the 'realism' and eating C-rations, low-crawling through two and a half feet of powder snow. (Just where was this jungle, anyway? The North Pole?) It was at this point that I accepted that my Army career was going to be truly bizarre, and decided to just roll with it, come what may.

The only thing different would be that there would be no Escape and Evasion problem. It's a bit hard to escape and evade if you leave footprints in the snow everywhere you go. I was scheduled for just three days, rather than ten. I qualified Expert with the M16. That made me feel better. Low crawling through two and a half feet of powder snow on the fire and maneuver course reduced me to giggles. It was catching, and soon all of us were laughing our heads off as the range sergeant yelled at us. He finally blew his whistle and dismissed us.

I was supposed to clear post the next day, but my name was not on that roster. When I mentioned that I was supposed to be done, the training sergeant looked at his clipboard and sneered at me.

"I don't see your name here. Off you go."

I was sure there had been a mistake, but had taken that notion as far as it could go. I went off to that day's training, on how to detect booby traps. I felt like a booby myself.

Two hours later, that same sergeant comes running up to the class, yelling at me for almost 'missing a movement' and hustles me back to the TTC assembly area to turn in my winter gear.

My old Commanding Officer from G Company, Captain Crybalski, was standing there, shaking his head. "Hamit," he said, "You'd fuck up a wet dream."

It was an unfair, hurtful, and very inappropriate comment. None of this was my fault, and the reply that sprang to my lips died unspoken. I was getting out of there. Why mess it up by expressing my heartfelt opinion that Crybalski was an asshole and a bad leader for saying such a thing?

I went silently, drew my tropical uniforms for Vietnam, and my pay and airline tickets, and left Fort Devens behind. Compared to this, Vietnam would be easy, I thought.

Little did I know.

My father's last letter received there had some heartfelt commentary about the war.

"I am sorry you have been alerted for an overseas assignment. With things the way they are, it will have to be almost certainly Vietnam or Korea. It seems a little ironical to me that while I was serving those unaccompanied tours to the Philippines, Japan, and Korea, and going through all the discomfort and loneliness, I used to try and console myself by thinking, and sometimes, even saying aloud, that by doing what I was doing, I was making it so that my son would never have to do what I was doing. Well, apparently it did not work out that way."

I had not yet told him that I'd held up my hand and volunteered. Anything to escape the morass of Code School. Popular culture had not yet turned Vietnam, with films and television about war, and especially that war, into a horror show. It was still a noble, heroic endeavor to me. I spent my boyhood getting ready to go fight for my country, somewhere, somehow, not realizing that almost none of it is what you see in the movies.

Hollywood does not make documentaries, even when they label them that. There is always a lie involved, and the lie I bought into defined me as a person and a man. I am one of those who run toward trouble when everyone else runs away. I am a Soldier, a word with a larger meaning than just being in the military. Were I not that way, I would have ignored the drugs that flowed into Iowa City and began destroying the lives of people I cared about.

Dad would hear none of this. I had learned long before not to trust him with anything really personal. I needed his love, not his critique, and that passage above showed that he did have some of that, but I was, and always would be, his dumb kid.

I was relieved that I would not have to go home on leave before going to the War. Things were already depressing enough. And I was actually looking forward to seeing what it was like. Pure intellectual curiosity, like an interested tourist.

Private Francis Hamit

CAN THO, SOUTH VIETNAM 1968-69

CHAPTER 11

A few days before I left for Vietnam, Walter Cronkite, the most trusted newsman in America, delivered his analysis of how the Vietnam War was going in the aftermath of the Tet Offensive on January 31. 'Uncle Walter' stepped over the line from reporter to seer, and declared it lost, despite the fact that the Viet Cong had had its collective ass kicked, and was pretty much out of business.

John Wadje, one of the Menehune platoon who actually was in Vietnam when the Tet Offensive occurred, sent me this about how resistant the Generals and higher ranking staff officers were to hard truths:

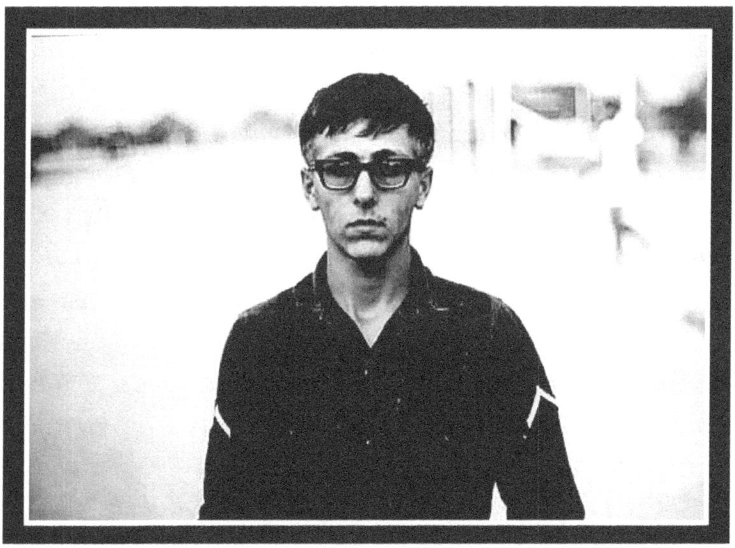

Private John Wadje

"Francis, In response to the question about the intel briefing to the 25th Div, this is the background. In September '67, I was suddenly called in from the field where I was setting up some antennas at Camp Rainer in Dau Tieng (Michelin Rubber Plantation). When I arrived back in base camp I was told to immediately report to the CO, Major Adams. Thinking that he has somehow managed to get me an R&R as I had been screwed out of mine in July and he said that he would find a way for me to get one before I left country. But that wasn't it at all, but was told by Major Adams that I was to be a presenter at an intelligence briefing for the division HQ. I didn't like the idea much but it did come with new boots, jungle fatigues,boonie hat, and a fresh haircut.

My role was to outline the progress by the VC in terms of communications equipment.

When I first arrived at the 372nd in January, 67, our Sig Maintenance OIC, WO Levi Joy, assigned me to analyze several captured VC transmitters. The first transmitter came directly from the Amateur Radio Handbook and was exactly like one that I had built as a high school freshman and I recognized it immediately. I then wrote a report on the testing that I did and included a circuit analysis and schematic.

Mr. Joy was impressed, so he gave me that job from that time on so that whenever I came in from the field, I would spend several days going through and analyzing the latest captured equipment. Now, this equipment progressed from crude home made radios in January constructed from scrounged garbage dump parts, to radios that were obviously assembled from new parts and competently soldered.

So, I was to present and explain this progress over that eight month time. But while we were in the briefing, I paid attention to the other presenters who were talking about the huge buildup of VC and NVA soldiers in War Zone C, especially along the Cambodian border at the end of the Ho Chi Minh trail.

They talked about the increased number of units that had come on line and were broadcasting with good stable transmitters. Likewise for such things as orders from their HQs for munitions, uniforms, rations, etc, as well as acknowledgments from those units when they received their supplies. They also noted that the munitions that they ordered and received were not being used but stored.

At the same time, they noted that the two major VC training units in the Iron Triangle were exceptionally active throughout the summer, especially the mortar and artillery units.

And, finally, they predicted a probable winter offensive, most likely at Tet when everyone would be celebrating, especially with fireworks.

But it was like talking to the wind, as every one of those officers present had their own ideas about the enemy and could not bring themselves to rationally hear what we had to say but instead seemed to spend their time talking to the General and expressing their own "informed" opinions so that it was like watching a circle jerk.

And they certainly disagreed with the idea that there were over 20k soldiers on our side and twice that many on the Cambodian side.

These briefings continued each month into December when the General decided to test our "theory" and installed an FSB near Katum at Soui Cut over the Christmas truce.

Major Adams and our operations officer objected to that idea, and told them that the NVA unit in charge was in command of a highly reputable colonel who could easily muster from 1,500 to 2,500 troops to attack us in a matter of a few days and that a FSB with only 1,500 troops simply wouldn't do it and would be overrun in a matter of minutes.

So, the only concession the HQ made was to add a 155 battery and another 100 or more soldiers, which Adams said was woefully inadequate as we were most likely to get attacked with an attempted overrun.

And that is what occurred on the night of January 1 & 2, 1968.

That the enemy could do it at all was the big shock, and went against all the happy-talk and self-aggrandizing BS that General William Westmoreland was peddling. Westy wanted a set-piece battle where the war could be decided by 'Real Men'. The sneaky VC refused to oblige him, and were fighting the real war in the streets of America with allies from the Children's Crusade of the anti-war movement.

As it turned out, the kids were right and everybody knew it. The radicalism spread from the high schools into the colleges and universities with concerned parents contacting the FBI and offering to spy on their own children just a decade past the excesses of the McCarthy era.

The high command knew it too. The US Army ran a war game in 1964 with General Earle C. Wheeler, the Army Chief of Staff, taking General Giap's role in a team of military and diplomatic experts.

The question was whether or not strategic bombing of North Vietnam would make Giap stop sending support to the Viet Cong. Would the threat to "bomb Hanoi back to the Stone Age" work? Tragically, the answer was no.

Even after the Viet Cong were virtually wiped out in the Tet Offensive, Giap had plenty of men to train and send south to fight, and women to fill in as truck drivers, medical personnel, scouts, spies and snipers. All highly motivated.

They knew the war could not be won and sent us anyway. Westmoreland looked like the popular culture model of a General. In retrospect, it looks like he was set up to fail.

North Vietnam was a Soviet client. The USSR backed their play with hard currency, weapons and gear, so much so that it contributed to its collapse two decades later. Perhaps that was the real game, using a hot war to win the Cold War. Was it worth all the casualties, turmoil and pain? It all followed that 1964 war game – almost to the letter.

Even after Tet, Westmoreland was on national TV, saying that the VC attacks were just a "feint" and the main attack from the north was still coming. Shortly after, that he was promoted to Army Chief of Staff and sent back to Washington. Not relieved of command, because of the optics; "kicked upstairs" to get rid of him.

In time, General Creighton Abrams would win the military war, and the VC and NVA would win the political one – the only one that counted: the one for 'hearts and minds'.

From where I was, it looked damned dangerous, and my family and friends thought the same. My now-retired father sent a letter warning me to be careful at Tan San Nuit because troops had been killed on their last day in-country by mortar attacks while waiting to board the plane home. Very encouraging.

I was apprehensive, but also felt challenged. There was so much BS in the media about the war, and so many rumors in the Army itself, that I had to go and see for myself. I'd already been in the Army for a year, and found what I'd seen disheartening. Going to the war was, I thought, my best chance at doing something meaningful in the Army.

ASA was still in shock from the 'Pueblo Incident' – the US Navy intelligence collection ship, with a crew of 83 had been captured by North Korean Navy patrol boats, and the realtime messages that appeared in our communications centers worldwide were traumatic. Despite our intense fetish for compartmentation and secrecy, details leaked out. That crew might be Navy, but we were all NSA, and a strong bond was present. The machine guns on the Pueblo's deck were so covered with sea ice that they could not be freed to defend the ship. A mountain of Top Secret material and some intact coding machines had been captured along with the crew. It was an intelligence disaster.

The response at Fort Devens was a notice posted on the bulletin boards ordering us not to discuss the matter with civilians. At first we were puzzled, because we'd never heard of the ship. Then the news broke in the media. There but for the grace of God.

I missed most of it because I was freezing my ass off at the Tactical Training Center doing my mandatory 'jungle training'.

Between planes in San Francisco, before boarding the contracted jet airliner that would carry me and several hundred other soldiers to Saigon, I met up with an old friend from high school. He was a year senior to me. We shared interests in firearms, military history, and mapping. We'd had long talks together as we hiked the Marin County hills, carrying our rifles for expeditions that included target shooting and unspoken fantasies about being infantry on patrol. We were always careful not to point our weapons in the wrong direction, and to be sure that our bullets would not travel beyond the target area.

It was four years since we'd seen each other, and we both were very surprised how much the other had changed. I, in my tropical tan Army uniform with MI brass, while he looked like the epitome of a hippie graduate student. He'd stayed local, and now attended San Francisco State where the Black Panthers and the anti-war movement were both in full cry for change. The radicalism also infected Tamalpais High School in Mill Valley, where we first met.

But he quickly revealed something we both had in common. We were both spies. Him more than me just then, because I'd washed out of my school. He was now a leader of a campus radical group, and San Francisco State was in constant turmoil.

"Don't they realize that you used to be a Young Republican?" I asked, amused.

He shrugged. "I told them I've seen the light. This war changed a lot of people. Why should I be different?"

"So you're against the war? Should I even be talking with you?"

"Relax. I'm working for the FBI. Your clearance is safe. In fact, they asked me about you." He described how he and another old friend had copied his group's entire membership list and passed it on. I was uncomfortable with that. I objected to political spying.

"This isn't McCarthyism," he said; "Most of these campus groups are harmless. They couldn't organize a keg party, much less a revolution. But there are some real nuts on the fringe. They can't wait to blow something up. If you were here, you'd be doing the same."

"I doubt it," I said; "I'm not that good an actor." I described some of the undercover narcotics work I'd done in Iowa City. "Never toked a joint or took a hit," I said, "And broke up with my fiancee over her use."

"How did you do it, then?"

"Just listened. People like to brag."

He nodded. "Everything's upside down. Dr. Hayakawa is not such a liberal anymore."

S. I. Hayakawa was a famous semanticist, the author of ***Language in Thought and Action***, and a tenured professor of English at San Francisco State. By the end of the year he would become its Acting President, in open conflict with various student radical groups, and eventually a Republican Senator from California. The liberalism he'd preached when his son Alan and I were best friends was fading fast.

"Sounds like a real balls-up over there," he added. The Tet Offensive was still reverberating on the nightly news. That iconic Eddie Adams photo of General Nguyen Ngoc Loan shooting the Viet Cong scout in the head was still everywhere. "Can't you get out of it?"

"Actually, I volunteered."

Why would I want to get out of it? I was a soldier and my country was at war. My place was in the combat zone. This was a minority opinion, even within the Army, as I was to find out.

A mutual friend had been an early volunteer and survived a year as a helicopter door-gunner, a job where the life-expectancy was usually measured in minutes. He'd become a great re-teller of improbable war stories, and seemed rather odd now. But it was what he'd gone looking for. I expected a quieter life. Not all that quiet, but not that crazy either.

"Well, be well," he said. "Semper Fi." Our mutual friend had been a Marine, and he simply didn't know the difference between that and the Army. We parted and I did not see him again for several decades. The friendship did not endure. I'm not sure why.

The long plane ride to Saigon was packed and uncomfortable, but preferable to the alternative of a three week vomit-filled sea voyage on an old freighter converted to a troop ship.

There were civilian stewardesses. Very pretty girls, very kind. One optimist several rows ahead to me asked one to marry him. She replied, "Well, I would, but I don't think my husband would like it." That got a big laugh and drew a line. She was not available.

It was a long flight. I read a book and slept as much as I could.

The heat and humidity that greeted us at Tan Son Nuit airfield was simply amazing. Coming off that airplane was like facing a blast furnace. A week before I had been freezing in woolens and mickey mouse boots, low-crawling through two and a half feet of powder snow. Now sweat instantly soaked my tans, starting with the armpits and my back.

We were herded off the plane and formed into ranks by the Major who'd taken charge of us during the trip, and some senior NCOs. I ended up in the rear rank, trying to balance my duffle bag in one hand and my black leather Valpak (a modern take on the carpet bag) in the other.

On a platform, a Sergeant First Class (E-7) stood with a clipboard in one hand and a microphone in the other. He was dressed in jungle fatigues, a new uniform available only in Vietnam. He looked cool and relaxed.

"Okay, listen up!" he said, his voice reverberating across the tarmac. "I will call out the units you are assigned to. Check your orders. If you are not assigned to a unit, then we will send you directly to 'the front!'

(Everyone was assigned. This was a device to make sure that men did not end up in the wrong unit and have to be transported again to their proper assignment. Also, there were some Sergeants lurking on the fringes of the crowd, hoping to shanghai unwary replacements into their shorthanded units – and most of those were at 'the front.' Some ASA soldiers had been diverted this way, despite the phrase in their orders that said they were 'under the control of the Commanding General, US Army Security Agency and may not be reassigned.')

His next command came as no surprise. "Anyone assigned to the 509th Group grab your gear and step forward."

There were three of us, all from Fort Devens and Company G. Slowly, wrestling with our baggage, we shambled forward.

"Come on, Superspooks! Get a move on."

So much for our cover.

Soon we were in the back of a 3/4 ton truck bouncing along the road until we got to the headquarters of the 509th Group (Radio Research). This was the umbrella organization for all ASA units in South Vietnam. The 'Radio Research' moniker was our 'cover.' Supposedly, there were sound geopolitical reasons for this gloss. It was allegedly a violation of international law to base ASA troops there, but the real reasons were more compelling. The Viet Cong had a bounty of 100,000 piasters out for any ASA soldier captured. They also had a standing offer of $10,000 US for any member of ASA who defected. (None ever did.)

The secrecy was necessary, because we had people assigned to every Army unit 'in-country', in units which ranged in size from small detachments to a full-blown field station. I am not going to recount them here, because there is an excellent book, **Unlikely Warriors: The Army Security Agency's Secret War in Vietnam 1961-1973 (2013)**, that will give the interested reader a pretty complete overview. There are some older books, now out of print, that are also informative, such as **The Most Secret War (2006)**, published by INSCOM and the still partially redacted US government history, **Spartans in Darkness (2002)**. I recommend all of them.

The cover designation helped us maintain a strict security discipline. While we were all members of the Regular Army, and all volunteers, we were also employees of the National Security Agency, and not responsive to local command and control. Our primary responsibility, from the first day in country to the last, was to not draw attention to ourselves or our unit.

How well did this work? Jerry Orlemann, who was a Traffic Analyst at Ben Hoa, relates this story:

"A friend, when I was awaiting orders in Saigon took me to a bar downtown. Some cute little thing came over to me and plopped herself down in my lap. We began talking – or at least she did – and the first thing out of her mouth was "You new guy. You ASA." I stuttered and she continued, "Everybody here ASA. This ASA bar." I stuttered some more and she hit me with: "You must be 98 Charlie (Traffic Analyst). They short 98 Charlie. You go to Bien Hoa. They short 98 Charlie at Bien Hoa." Three days later I got orders for Bien Hoa."

Later that day, I was issued jungle fatigues. "They're supposed to be loose," the supply sergeant said, "for ventilation. Don't get them tailored. You get an Article 15 if you do. Don't let anyone tell you that they need to look 'STRAC'. That's stateside bullshit. This is 'Nam. Things are different here."

Truer words were never said.

We would not be issued a weapon until we got to our unit, and that made some people very nervous. Yes, we were in the middle of a big base, surrounded by thousands of our fellow soldiers who did have weapons, but the VC had just proven their ability to strike anywhere, anytime. They owned the night, and had a second mortgage on the daytime, too.

Between the news coverage and the war stories, we 'newbies' all knew that we had landed in deep shit. I never felt more naked than I did walking alone to the mess hall that day in my new jungle fatigues. I'd fired 'Expert' with the M16, but it would be hand-to-hand if a wily VC scout sprung out of the grass and attacked me.

The other thing was the smell. War is an industrial process, and Tan San Nuit smelled of oil and solvents and dust and sweat, but the overlying (and overwhelming) odor was shit. That smell was everywhere because this was a third-world country where human feces were the only fertilizer available. And rather than contribute this valuable commodity to the locals, the American forces burned it. More sanitary, but it created its own unique form of air pollution: a black miasma of smoke that drifted everywhere and got into everything.

CHAPTER 12

Soon I was shuffled off to the Headquarters of the 224th Aviation Battalion (Radio Research). ASA had its own air force! It was a quick in-and-out as I was reassigned to a subordinate unit, the 156th Aviation Company (Radio Research) at Can Tho Army Airfield in the Mekong Delta.

It was less than a hundred miles away, but we would have to fly there. The roads were not safe, even in daytime, and ASA personnel were not supposed to use ground travel in Vietnam anyway. The government had a considerable investment to protect, for security clearances and training. Operations might take us in harm's way, but minor administrative chores would not.

Three of us joined the passengers on a C-130 transport aircraft going to Ben Tuey, a big Air Force base north of Can Tho. It was a very uncomfortable ride, because the few canvas seats on the sides went to higher ranking officers, and the rest of us were strapped to the deck where we felt every jolt. The engine noise was deafening as well.

We got out of the plane and walked around, trying to restore circulation to our abused bodies. The heat and humidity were oppressive in the extreme, and there was no water available.

I felt sick and disoriented, and noticed two new Huey Cobra gunships sitting at the side of the runway. They were painted with Air America colors and logos, and shark's teeth decals on the front. They looked like giant bugs out of a bad science fiction movie.

"That's one hell of a civilian airline," I said to myself. Of course, everyone knew it was a CIA front, but seeing the most advanced air-attack helicopter available, in mock civilian dress, was just too bizarre. Surely I was hallucinating? (Apparently not. A source recently confirmed that the Huey Cobra was issued to the CIA even before the Army and Marines.)

Then it was back on the plane for a ten minute jump to Can Tho, where we reported to the Orderly Room for the 156th, and were told to go to the barracks, find a locker and a bunk, and get our gear squared away, and then report back.

The barracks were two-story temporary buildings with the sideboards tilted at a 45-degree angle to provide ventilation, the gaps covered with wire screens. They were shells with no insulation, and the floors were unfinished plywood. There were small jagged holes in the boards and the screens.

"Shrapnel," said a guy who was sleeping in the middle of the day, having pulled guard duty the night before; "From the mortar attacks."

"That happen often?"

"Only about two or three times a week." He shrugged. "You just run to the bunker and wait for it to be over."

Easier said than done. The current bunker was a steel shipping container covered with plywood and sandbags. The company was originally a detachment of about 20 with a single aircraft, and expanded too fast to meet customer demands for more SIGINT. Now it had six aircraft and a roster of 167. The bunker was too small, and the steel container did not belong to the Army but to a contractor, and they wanted it back. Moreover, a direct hit would shower additional steel fragments from the container on the hapless inhabitants. It was a quick and lazy way to build a bunker. Field improvisation in the wake of the Tet Offensive attacks.

Can Tho barracks

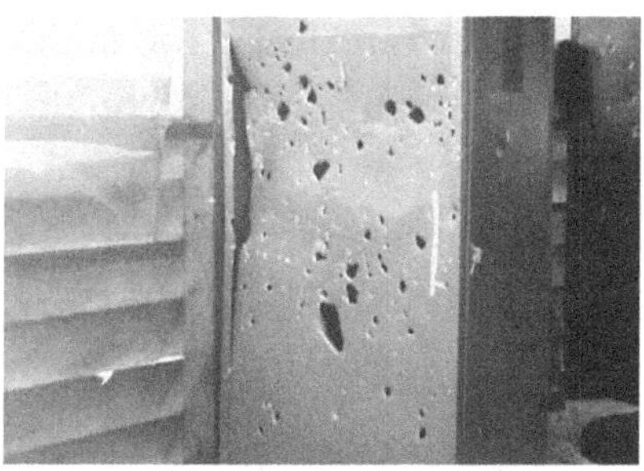

Shrapnel damage photos

Two new ones were being built with sandbags over a wooden frame outside our barracks, which were part of a block of 32 identical structures. We were hiding in plain sight. There was no exterior signage that identified our barracks as ours. We had no mess hall of our own, but used a common one belonging to the 13th Aviation Brigade. Our circular, unofficial, unit patch, a very colorful one with a tiger's paw imposed on top of a crude map of the Mekong Delta and lightning bolts at each side with the legend 'Delta Stalking Grounds', was encased in a plastic holder to be buttoned on when we were at work or other duty places on the base, but otherwise left in our locker. Never to be worn off base for security reasons. Or when we pulled guard duty on the perimeter. Since there were so many new people coming in, and others leaving after their one year tour was finished, it gave us a quick and easy way to recognize each other, and reason to challenge people from other units who wandered onto our turf uninvited. We had a platoon of ASA MPs, but security was everyone's job.

The first few days were spent filling sandbags and using them to build the new bunkers. In the extreme heat, it was man-killing work. This was to acclimate us to the hot humid weather. It was 110 in the shade, and there was no shade. We had a mortar attack the first night I was there. I recall hearing a soft popping sound, a shout of "That's it!", a crashing sound as forty pairs of feet hit the plywood floor, and a rumble as we all ran down the stairs, after which I woke up, running for my life, as we all crowded into the still unfinished new bunker, and hugged the ground. There was some praying going on. The 'all-clear' sounded half an hour later. Just a little reminder from Mr. Charles that he was out there. Welcome to the neighborhood.

Since ASA in the field worked rotating shifts, there was no morning formation. You took a shower, if they were working, put on your uniform, ate breakfast and walked to your assigned work area. No marching in formation. Walk to work, and be on time, like any other adult. The Orderly Room was on the second floor of another barracks building, but the C.O.'s office was in Flight Operations, which was next to the Flight Line where our aircraft were. Across the road was a nondescript Quonset hut that housed the Operations Center, a highly restricted facility behind a chain link fence and barbed wire; there was an MP building where everyone had to check in. Only those granted access were permitted inside, and their Worldwide badges were kept in a rack by the door, with a double guard post. One inside and one outside. I never saw the inside of the place the whole year I was there. I had no 'need to know'. There was a second Operations Center in the officers' hotel downtown, used for mission planning. I saw that once.

Bunker

HQ, 156th Avn Co (RR)

The flight operations building was metal. It was between our maintenance area and the runway. The runway was made of Marston mats, metal units that were laid down over packed river sand, and they vibrated every time an aircraft landed. Besides the aircraft and helicopters overhead, and the constant landings, there was also occasional automatic cannon and rocket fire from the Huey Cobras that flew overhead to protect the base around the clock. It was a vibration soup. One that has side effects, decades later, as bad as Agent Orange.

We were each issued an M16, along with ten clips of ammunition in a canvas bandolier, and told to clean it daily. There was fine sand blowing everywhere, and it could foul the mechanism, even when you kept the rifle in your locker. Unless on Alert or guard duty, the weapon was never loaded. It had a hair trigger. We went to the range one time to learn how to fire it on full auto. The preference was for short bursts, since a 20 round clip went pretty fast. That trip to the range would be the only time I would actually fire this, or any other weapon, the entire year I was there.

CHAPTER 13

The base had a big club for enlisted men, and an officers club used by other units, but our pilots did their drinking at the hotel, and the senior NCOs had their own little room someplace where, as First Sergeant Ezio Tozzi later told me, they had every kind of liquor. Tozzi didn't drink, and was appalled when he joined the company in May at how many of his subordinate Regular Army career NCOs were obvious alcoholics. Most of them were not ASA, but had been levied in from other units.

It was here that I first saw the damage done to ASA by the rapid expansion to meet the SIGINT demands of the Vietnam War. Our strength was about 30,000 worldwide, and 6,000 were in Vietnam. The turnover in the lower ranks was intense. One-year tours and four-year enlistments. Reenlistment rates were the lowest in the Army. Really talented operators moved onto better paying civilian careers with NSA, CIA, or other intelligence agencies. Slots among the cadre often went unfilled.

The Regular Army provided support personnel replacements. NCOs who were sub-par and also short of a graceful retirement were put on these levies by their former units, and sent to our units in the field to fill non-intelligence slots in maintenance, supply, aviation, motor pool, and so on. Other units viewed the levies as a golden opportunity to dump their trash on us. While it was increasingly harder to get people to reenlist, it was even more so to discharge ineffective and alcoholic NCOs, who wanted to stay in the service until they could get their 30-year pensions. The 'E-5 over 20' in my company at Fort Devens was a case in point.

Operational and communications, and some high tech maintenance jobs, and the MPs were ASA, and then there were the school wash-outs like me, who had the ASA clearances and could train 'on-the-job' (OJT) to fill the many clerical positions required by the company's Table of Organization and Equipment. Very few of the pilots were also MI officers. Most were Captains, Lieutenants and Warrants who'd graduated from the fixed wing aviation schools. Typical flyboys who liked to drink and carouse, and hated paperwork. That was unfortunate, because many of them were assigned as the officer-in-charge (OIC) over one or more of the functions that the Army deemed essential to good order and discipline, even in a combat zone. NCOs ran these sections, but a commissioned officer had to sign the paperwork. The Operations Officer was from MI, and not a pilot. Neither was the Property Book Officer (a Warrant), but everyone else, especially the Commanding Officer, flew missions daily. We were an airborne radio direction company, tasked with finding Viet Cong units.

The first official combat casualty of the war was James T. Davis in 1961, an ASA soldier. He'd been with a mobile Vietnamese radio unit, training them on RDF procedures, when they were ambushed and wiped out by the VC. This became part of the mythology of ASA in Vietnam, that we were at high risk. Putting the RDF in airplanes was a safe alternative, and more efficient, because it lessened interference from the jungle overgrowth and other terrain features. It was a lot harder (but not impossible) to ambush an airplane. I knew none of this then – I was a 'newbie', a babe in the woods.

* * *

After three days of filling sand bags, I was sent to fill an open slot in Flight Operations, and to train on that MOS, 71P. Because I had passed the Army typing test at Devens, I was designated a

71B, or ordinary clerk-typist. There were a number of unfilled slots in the company that I was tasked to cover. The structure was designed for a stateside unit. You would have thought that having a training and education NCO was not important in a war zone, since we were training every day just by being there, but we were held to the same standards as every other unit in the Army worldwide. Someone had to do these little jobs, or we got low marks for our 'readiness'.

Requirements that seemed logical and proper stateside became burdensome, or were ignored, in the combat zone. Higher headquarters demanded compliance. That we weren't in Kansas (a stateside base) anymore, was hard to get across, Officers and career NCOs coming in from Regular Army units in the USA, or a European post, experienced culture shock at how 'slack' we were. They wanted us 'STRAC' (strategic), with everyone is crisp neat uniforms and everything not just according to the ARs but to their exacting and very competitive standards.

They wanted to impress the brass with how well we looked, rather than how well we performed.

* * *

A Regular Army E-7 reported into the 156th, and was horrified to find that everyone kept their M16s and ammunition in their lockers. He loudly demanded that everything be immediately returned to an arms room – we didn't have.

We had a closet in the Motor Pool where we kept some M60 machine guns and M79 grenade launchers that were not on our TO&E, and other gear that wasn't really ours. That was it.

The idea was ridiculous on its face. If there was another attack, there would be no time for people to line up and sign out their weapons. And this was ASA, where people were trusted once they got their clearances. His problem was one with which we were already familiar. He'd been in the Army 17 years – and in a combat zone, for the first time, two days.

He was also harassing junior enlisted men over their 'sloppy' uniforms. Jungle fatigues, you will recall, were designed to be worn loose for ventilation. Then everyone lost weight because of the miserable heat and humidity. After his first mortar attack and Alert, he became a more reasonable man, but it took awhile to boil the Regular Army out of him.

* * *

Our primary mission aircraft in the Delta was the U-6 DeHaviland Beaver, a very stable single engine platform that looked sort of like Charles Lindbergh's 'Spirit of St. Louis' (with better windows in front, and a lot of antennas on the wings). The plane itself was an antenna. The 40 foot wing was designed so. It held a pilot, co-pilot and an operator who sat in the back and worked the radios. It was the operator who controlled the mission. The pilots, as one of them complained to me, were just 'driving the bus'.

But I knew none of this first hand. There was one U-6 reserved for administrative duties, and I would occasionally be detailed as a courier, and be given custody of several locked mail bags to be flown up to Saigon on that aircraft. I was the temporary custodian of 800 pounds of Top Secret paper, armed with my M16, my mail clerk's .45, and two thermite grenades with which I was supposed to burn it all if the plane went down.

Most of my duty time was spent in the office. The NCOIC of Flight Operations was an irascible hillbilly, who was also the acting First Sergeant until Ezio Tozzi transferred in. He hated 'college boys' and was hard to please and not much 'hands-on', preferring to throw the relevant Army Regulation on my desk and tell me to figure it out for myself. I got all the details he didn't want to handle. Education and Training, and Awards and Decorations, were my primary jobs. He never trained me in the 71P Flight Operations MOS.

I was eager to redeem myself for my failure at Fort Devens, so I took them, and also became the mail clerk. The mail clerk was issued an Army Colt .45 pistol because he sometimes handled money, and the fact that I had it, made me one of the three classified couriers for Top Secret/Crypto material. I did not get the usual military holster with the flap to hang from my web belt. I bought a shoulder holster on the economy, which was much more convenient to wear and made me look dangerous. I would have actually preferred one of the .38 caliber Smith and Wesson revolvers issued to the pilots and operators, who wore them low slung, gunfighter style, but was told that regulations required I be given the .45. I carried it outside the office, and took it off, along with my baseball cap, when I was typing or doing other office work. The holster strap was uncomfortable, restricting movement.

When we got a 16mm Bell and Howell projector in for showing training films, I had to run that, too. Fortunately, this was a skill set I'd acquired in high school. The bonus there was that we also got special 16mm prints of first-run Hollywood movies, and black and white prints of two new television series: "Star Trek" and "Mission: Impossible".

I was happy to run the movies at night. It gave me status, and, aside from drinking, there wasn't that much else to do other than the occasional game of poker. Reading was hard because the lights were too dim.

"Star Trek" was a military we would have all loved to have served in, and "Mission: Impossible" was what we foolishly and falsely thought we were signing up for when we selected a career in Military Intelligence. It gave us a lot of laughs.

CHAPTER 14

I read Army Regulations in my spare time at the office. It wasn't that hard, and I sometimes had lots of time to fill. I had to pick up the mail every day and distribute it, and when the pilots returned from their four-hour missions, they wanted to escape back to their hotel downtown and drink, rather than do their other jobs. I was asked to drive them sometimes, but since night-time light discipline (to prevent making us a target for the VC mortars) kept us out of the office after sundown, I came to resent the time away from the clerical chores my NCOIC tasked me with. He would chew me out if they were behind time, and every drive was an hour or more out the office. Lots of typing, which I did not do well unless I went very slowly. If I rushed it, I had to do it over.

We had a brand new M151 quarter-ton vehicle (jeep), and that model was notoriously unstable in the best of circumstances. I drove fast because I had work to do. Over 20 MPH it became a bit of a thrill ride for passengers since the road to the hotel was densely packed with Vietnamese military and civilian traffic, as well as our own. Some of my passengers didn't care for my cowboy style, and would ask other enlisted men in Flight Ops to chauffeur them instead. At least one pilot said that riding with me was the most frightening experience he had in Vietnam.

1968 was already a weird year because of the Tet Offensive and the Pueblo crisis. Added to that was my father's decision to retire from the Army Medical Corps. He was walking away from a brilliant career to go to work for a big drug company and make some money. He was on the short list to become the next Army Surgeon General, something I'd found out at Fort Devens when an FBI agent came to query me about his politics.

"What are you talking about?" I replied. "He's a career Army officer. They don't have political opinions. He doesn't even vote." It was this FBI agent who told me that he was about to be promoted to Brigadier General, and that required a new security clearance.

When I asked Dad about it, he pooh-poohed the idea, and said that he didn't want the job. He had to fight to retire. The Board denied it, and the then Army Surgeon General had him in for interviews three times before he finally was able to become a civilian. He would have none of it. He had a sick wife, a daughter in college, and wanted to make some money for the first time in his life. The pay of a Medical Corps full colonel was $18,000 a year. The new employer tripled it. He suddenly had stock options, and a new Buick procured for him by the company at a bargain price.

"Generals don't get to do surgery," he said. "They get to do paperwork. I'm a surgeon."

All well and good, except that the new job was all paperwork, conferences and meetings. It was for a big drug company, doing what he'd done for the Army Surgeon General ten years before: administering research contracts and grants. Going in, he was enthusiastic about the job, but that diminished quickly. He really was a surgeon, and would take a big cut in pay 18 months later to return to the operating room.

He was also the Army's Medical Corps Inspector General in Vietnam in November 1967. He was not impressed with this war, and didn't think we could win it. And said so openly, not just aloud, but in his trip report. He returned from Vietnam with three recommendations for courts martial and two for Soldier's Medals. Those last were for a surgeon and a technician who'd done a new surgical procedure, the 'grenadectomy.' They had removed a live 79mm grenade from a soldier's leg. If it had exploded, they would have lost their hands and eyesight despite the sandbag wall erected around the operating table. A senior E.O.D. (Explosive Ordinance Disposal) Sergeant stood by to take the explosive away. It was a near thing, and the surgeon who did the operation was the medical unit's former Commanding Officer, whom my father relieved and replaced on the spot, because he was a drunk who slept with his nurses. He would be court martialed for 'conduct unbecoming', but he would also get that medal. My father was hard, but fair. He'd loved the Army, but wanted no part of this war. Nor did he want me there, if you read between the lines.

"We have been watching the war news rather closely," he wrote me on March 3; "But it is hard to tell what is really going on, even when it is brought live and in color right into one's living room. From experience I know that the news hawks can't and don't cover all the action, and I know that many times things seem a lot worse than they really are to those of us back here than they are to most people who are there. It really is a crazily-run war."

He added, "We haven't had a really outstanding student riot for almost a week." He was now in a rented house in the Chicago suburbs and the run-up to 'The Days of Rage' during the Democratic National Convention six months later was already in preparation.

Before that Convention, there were three very shocking events. Lyndon Johnson announced that he would not run for another term as President. Martin Luther King was assassinated. Robert Kennedy was assassinated.

Each of these events, and especially the last two, were body blows to the soldiers in Vietnam. It felt like the country was coming apart behind us. The King assassination produced riots nationwide. The one of Robert Kennedy guaranteed a

Republican victory in the fall election, and no one wanted Nixon as President. No one I knew there, anyway.

"What the hell is going on back there?" we asked each other, more than once. But we had more immediate problems in front of us. It was, as I said, an industrial environment, so there were all kinds of ways you could die, even without combat. I had never driven a stick shift, and had to get a military drivers license. I had five minutes instruction on a two and half ton truck, disgusting the Motor Pool Sergeant with my ineptitude. He said, "You're getting a brand new jeep in your section. You better practice some more. Damn thing has a Corvair suspension, and will flip at six miles an hour if you turn too sharply." The Corvair was the car that made Ralph Nader famous when he wrote his expose ***Unsafe at Any Speed***.

CHAPTER 15

The airfield was attacked during the Tet Offensive, and parts of Can Tho destroyed by pitched battles between the ARVN and the VC. From the war stories going around, you would have thought it was the Battle for the Alamo.

John Reid , one of our operators who did two tours at the156th, posted on Facebook about his experience during the Tet Offensive:

Remembering Tet after 56 years.
by John A Reid

May 24, 2024. It's 10:20am and I'm sitting in the dermatologist exam chair in my underwear, waiting for the dermatologist to examine my body. My mind starts to wander back to Tet. I suppose that doesn't mean much to anybody not affected by that Lunar Holiday back in 1968 in Vietnam.

January 30, 1968. That's the date the North Vietnamese and Viet Cong started their offensive in South Vietnam. Could have been as early as the night of January 29/30. Some units in the northern areas of S. Vietnam jumped off early. The objective, I've read, was to take control of the provincial towns and cities and gain popular support for a Communist government.

The morning of January 30 I woke up and got dressed in my fatigues. I was assigned to the 156th Avn Co (RR), Can Tho, South Vietnam, on the 1st of January 1968 as an Airborne Radio Direction Finder operator. It was my day off. I had extended my service in S Vietnam for an additional 12 months. As an incentive to extend, the Army gave a 30 day special leave. I had decided to take the leave and be the best man at my brothers wedding. I was scheduled on a flight out of Tan Son Nhut, Saigon, the next afternoon. I had an early morning flight from Can Tho to Saigon the next morning to make the connection.

I hitched a ride from one of numerous jeeps running between Can Tho city and the Can Tho Army airfield which was just a few miles away. When I got to the barracks it was kind of empty. I talked with a friend of mine. Reggie told me they had evacuated all the aircraft with crews including the DF operators to Thailand. A major attack was expected. We were both DF operators and I had noticed on my missions that the VC units we were tracking had been moving closer to population centers, including ours. So it was no surprise to us that there was an impending attack. I packed my small bag and got out my khaki uniform and got ready to leave in the morning. Really unconcerned with the VC. Later in the afternoon we iced down some beer and sat around bsing. Seems like about 10 or 11 that night we started hearing automatic weapons gunfire. We grabbed our rifles and headed towards the bunker out side our barracks. As we lay in the bunker facing out we watched green automatic weapons tracer fire between our barracks and the next barracks behind us. Green tracers are used by the VC. So we lay there getting ready for them to show up between the barracks. Before that happened two guys run up from our right with an M60 machine gun and set up between the barracks and start returning fire. That lasted for about a minute and the machine gun crew picked up the M60 and disappeared between the barracks towards our left. A Staff Sergeant comes up to us and orders a couple of other guys in the bunker to follow him and tells us to wait there as reinforcements, if needed. He disappeared around the end of the barracks along with the other two soldiers who had been in the bunker with us.

We waited at the bunker for an hour and then we were ordered to reinforce our line along the runway. We moved out with a Staff Sergeant to the runway and took up a position about mid-field and watched and listened. We didn't take any enemy fire at that position for the remainder of the night. There was plenty of small arms fire coming from northwest of our position where the jungle and brush overgrowth came close to the perimeter there. As it started to get light, the small arms fire ceased and it got quiet. At about this time we were told that some ARVN troops would be clearing the field in front of us. Shortly after that I watched two or three ARVN troops start moving through the field from my left to my right. About 100 yards out in front of us one of the ARVN troops reached down and pulled a VC body up out of the field to show us. He put the body down and walked further to the right and pulled another body up. He put that body down and continued to the right checking the area. At about 6am we were relieved from our position and told we could go back to the barracks and get some rest. I walked to my right to the end of the runway. At the end there was a defensive position and about 10 feet out in front of it was a dead VC. 1/2 of his head was blown off and he had a very grim look on his face. They got close. Then I headed back towards my barracks. On the way I walked past four VC bodies lined up next to a taxiway. I'd heard that 2 GI's had been killed during the night. In later reports I read that 69 VC and 5 GI's had died. I stopped by the CO's office and checked in. They told me I would not be leaving that day and that they didn't have any communication with any unit outside of the airfield so my leave was delayed. They also informed me that I had perimeter guard duty that night. As I was walking back to my barracks to get some rest, two U S Air Force F4's flew low down our runway and rocked their wings. Then pulled up and did a barrel roll as they flew away. Apparently satisfied we were still there. They thought we'd been overrun.

I managed to get a few hours rest prior to reporting for perimeter guard duty at the CO's office about 5 pm. I brought my M2 Carbine with as much 30 caliber ammunition as I could carry. Flak jacket and steel helmet. There were three of us. I was a SP5 and the other two were PFC's who were avionics guys. We headed over to base operations. Walking in file I was in the middle following the lead guy by about 5 feet. About a block from base ops office there was an explosion about 10 feet in front of the guy in front of me. We just stopped. No idea what happened. We were standing in the middle of the main base road. I thought, did the guy drop a grenade. Then another explosion happened behind the guy behind me about 25 feet away. Then it dawned on us. We were under a mortar attack. We headed for the ditch on the side of the road to take cover. We watched the mortar rounds land behind us. The VC walked the mortar rounds right into the barracks area. As we watched a GI came out of the second story door on the barracks and stood on the step platform just as a mortar round hit the barracks behind him about ten feet. He went down. I remember his legs dangling over the edge of the platform. We stayed put until the attack stopped. There were plenty of people taking care of the downed GI so we continued to base ops. Can Tho Army Airfield did not have any way to counter the mortar fire. The base was supposed to to be secured by the ARVN. Army of the Republic of Vietnam.

We got to the base commander' office and reported in. The Staff Sergeant in charge lined us up for inspection. There were about forty two of us. He stopped in front of me when he saw the old M2 carbine. He asked me how much ammunition I had. I told him 500 rounds. He was satisfied with that. Then the group was broken down into two groups. I was put in charge of one group and a SP4 was put in charge of the second group. The Sergeant explained that we would have two hour watches. Starting at about 6 pm. We would be covering the same part of the perimeter that had been overrun the previous night. Group 1 would start at 6 and Group 2 would relieve us at 8 pm. We would rotate that way throughout the night. He explained there was a listening post about twenty to forty feet out in front of the perimeter. The area out in front of the perimeter was a free fire zone. No permissions needed to fire. He also mentioned there would be a very small slice of moonlight that might be available, but there were clouds, so be vigilant.

Group 2 went to the guard duty ready room. There were bunks and chairs there where they could rest and be ready to go. They would relax until 8 and then load up on a deuce and a half truck and go out to the perimeter. We then loaded up on the deuce and a half and went out to the perimeter. Then we went to each position along the line as my guys set up a defense. After we were set up I went back up the line and talked a little with each guy in the group. 17 to 19 year olds and scared. They were mechanics, radio guys, and drivers, not riflemen.

Guard Mount, at 156th Avn Co (RR)

Two guys at each spot. I'd be out on the listening post with one of the guys. I told them all. There's a free fire zone in front of us. If you hear something, shoot it. There aren't any good guys in front of us. I made sure they understood. It was going to get dark. Very dark. I also made sure everyone knew the sign and countersign. I didn't want any problems or misunderstandings. Everybody was going to walk out in the morning. Nothing happened and we were relieved at 8pm. Back to the ready room for two hours. I don't remember anyone sleeping. Then 10 pm back out to relieve group two. If something is going to happen it will be between 12 and 2. Everyone is tense. It's dark. I'm not in the listening post this time. I'm moving up and down the line checking my guys. They're nervous. I remind them. Any noise out front, shoot. I'm not exactly relaxed. 22 years old, leading these 20 guys. Like they say, everything was fine, until it wasn't. About 11:30pm automatic weapons fire. Then pretty quickly the entire line opens up. I move back down the line to my most outward position. There was about a 6 foot gap in the defensive berm I had to cross. I stopped and then dashed across to the line's last position. The jungle came to within about a 100 feet of the berm here. The guys said I was pretty lucky. They said rifle fire followed me across the opening. I hadn't seen a thing. They were fine, they were returning fire at flashes out front that they said they saw. I next crawled out to the listening post. Made sure they knew it was me. They said they'd heard something and opened fire. I said ok. Hang in, nobody is going to get close. Then I crawled back behind the berm and went back out to my first position where the relief truck was pulling up. I told the SP4 in charge of the 2nd group that we were all going to stay here on the line until whatever was going on was over. We reinforced the line with his guys. I told the Staff Sergeant on the radio what I was doing. That we were taking fire. He had no problem with me keeping everyone on the perimeter.

We kept getting sporadic fire for most of the night until about 4:30am. I was on top of the berm watching and returning fire. I heard a helicopter start up behind me a couple hundred yards. Didn't think much of it. He took off away from us and gained some altitude and circled back. I heard him speeding up on an attack run behind me. Then all I remember is an explosion and flying upside down and landing on my back looking up as the helicopter flew over me. For some reason, he had fired a 2.75 inch rocket that had landed below me into the berm. As I watched the helicopter circle around, I located my rifle and helmet. The helicopter started another run and opened up with his mini gun and flew out in front of our position. I can't tell you how relieved I was. It's no fun staring down a Cobra.

Huey Cobra in flight

The incoming fire died off after that. I got on the radio and asked why the helicopter fired the rocket. I was told his aiming device was off. About as close as I've come to buying the farm. When the sun came up the ARVN came out. We were relieved, and left the position. On my way back to my barracks I talked to the Staff Sergeant. He said that they sent the Cobra up because they could hear rifle fire that didn't sound like most of the rest. I asked him if it could've been my M2? He said maybe. An 'ah ha' moment for me. I was also told later that the ARVN found blood traces in front of our position.

I went back to my barracks and got out of the dirty stinking fatigues from two days of crawling in the dirt. Took a shower and felt better. I went to the CO's office to see what was going on with my leave. I was told they needed to run a slick up to Tan Son Nhut at about 1, and I could ride on it. A slick was an aircraft with no mission gear. Made me a little happier and I decided to go to the mess hall and have breakfast. As I was walking down the road to the mess hall the base got mortared. The attack was not very far from the last place mortars were dropping. I watched as a pallet of

PX soft drinks blow up. Then the mortar adjusted and they started landing in the mess hall. I turned around and went the other way. Went to my barracks and changed into my khaki uniform. At 1 pm I was sitting in the back seat of an Army Beaver aircraft headed to Saigon.

At Davis Station, which is at the Tan Son Nhut airport, I reported in to the 244th Avn Bn (RR). They told me there was no commercial aircraft out of Vietnam at that time The airport was still receiving sporadic incoming fire, but they could get me on a C141 leaving in about an hour. I told them that worked for me. They drove me over to the Air Force parking area and dropped me off at operations. I checked in. They told me I was going to be an escort for KIA remains being returned to the U S. I boarded the aircraft with two other soldiers. There were about 12 aluminum coffins draped with flags. It was kind of solemn. We were out of Vietnam by about 4 pm. We landed at Clark AFB in the Philippines for fuel, and departed about an hour later for Elmendorf AFB in Alaska. We had to deplane there for the fueling. It was cold. The three of us were in our summer uniforms. The Air Force loaned each of us a blanket and sent us to a passenger waiting area. There was a pay phone and I called home collect. My mother answered and I told her where I was and that I would be home hopefully the next day. We said goodbye and it was time to board. We left Elmendorf AFB for Dover AFB, our final stop. Then I'd have to find transportation to Detroit. En route the aircraft had some sort of hydraulic problem and diverted to Kelly AFB where I was able to catch a commercial flight from San Antonio to Detroit. One day late for the wedding, but glad to be home. In 30 days I had to go back.

A sample of a soldier's life back then. Not looking for a welcome home or thanks for your service. Just trying to share what a soldier's life was like back then. So you might have some appreciation for what they go through. We're not perfect, but we step up when needed.

Admin. Beaver in the air

This is the most authentic account of those events I've found.

John is not someone I recall from my own tour, but the operators pretty much kept to themselves for security reasons. There were 167 guys in the 156th. More if you count the constant turn-over from the one year tour. Guys were always arriving to replace those going home or to another post. Junior enlisted men adjusted fairly quickly. Senior NCOs with no real combat experience, not so much. They stank of fear, and drank too much.

Some of our senior sergeants would yell at guard mount about staying alert on the perimeter because the VC were still out there and more powerful than ever.

Another unit salvaged a 20mm minigun cannon from a disabled Huey Cobra and mounted it on a jeep along with a 'Honeywell' or automatic 40mm grenade launcher. They called it 'The War Wagon' after a John Wayne film. It was used during Tet on the perimeter with great effect, sending streams of exploding shells into the center of the VC attack. A great example of field improvisation, but now, more than a month later, it was still being paraded every night because its creators were very proud of it.

A visiting Sergeant Major spotted it one day and went over to inspect it.

"What does this do?" he asked as he stood in front of the weird looking contraption and spun the barrels, shooting himself eight times in the abdomen and dying on the spot.

There were so many unique ways to die in Vietnam. The War Wagon was dismantled the next day.

Faux heroism from our senior NCOs could be tolerated. We were new guys. What did we know about what happened during Tet? We were already getting sideways glances and muttered comments about how 'shaky' we looked from the very NCOs who were supposed to lead and inspire us in the next attack. I resented this a great deal. I'd grown up in the Army. But I'd also been in long enough to know that I did not know it all, and that my best path was to keep my mouth shut.

The Awards and Decorations detail had been managed by someone long gone, and there was a pile of files to be processed. We generated so many Air Medals for our pilots and crews that subsequent awards were represented not by oak leaf clusters, but little brass numerals. (The record for our unit was 53 awards of the Air Medal in one year.) But these files were recommendations for awards for combat, earned for the defense of the airfield during the Tet Offensive – Silver and

Bronze Stars, and a Distinguished Service Cross, among some Army Commendation Medals with V.

All were correctly typed and carefully documented. Signed affidavits supported the actions described. I read them several times after one of those recommended stopped by to harass me for not having sent them to our headquarters. They only required my OIC's signature. I was to obtain this forthwith, and send them on.

The War Wagon, three views

My OIC was one of the pilots who hated paperwork. Not from the trade school on the Hudson, West Point, but a ROTC officer from CCNY, the same school that also produced Colin Powell. He was not one of the drunks, and openly criticized me for my adventures in Ben Xe Moi, but he had a wife at home and I did not, and I politely told him to mind his own business.

He was serious about the Army and already a Captain, but he grumbled when I insisted that he read every citation himself, and in detail, before signing them. Clerks were supposed to be self-supervising. But I had not written or typed these recommendations, and I'd noticed that the affidavits were mutually supporting; that the soldier recommended for one award would be supported by those recommended for others. The whole thing stank like a shit-burner. I was a newbie, an E-3, and a guy who'd fucked up his last assignment, so just saying so was not going to cut it. I had no credibility.

The Captain grumbled some more, like a kid wanting to go out and play rather than do his homework. Then he caught wise that something else was up.

"What's going on, Hamit? I know you. You're up to something."

"Please, sir. Just read them. Think of it as a field problem."

He did so, becoming more and more interested, and not a little impressed. This was stirring stuff. Real heroics in the war, that our unit would never see again. He got to the bottom of the pile, and asked, "Hamit, where are the Purple Hearts?"

"Sir. There are no Purple Hearts." Despite language in those documents such as 'despite his wounds, he continued to advance' there was not a single Purple Heart form.

His face flushed red as he realized that he'd been reading fiction. I'd checked around, and found that the events described had never happened. None of the phoney citations were for our MPs, who had been on our assigned part of the perimeter and part of the reaction force. These guys were back at the barracks as a reserve force.

The main VC attack had been a pathetic farce perpetrated by the local VC cadre, who rounded up some teen-aged boys, forced obsolete Chinese SKS carbines into their hands, and drove them into the meat grinder of our machine guns and Claymore mines. It was a slaughter.

I was disgusted that allegedly professional NCOs would sully themselves and the honor of our unit with such a dubious scheme. It would have never worked. Awards for combat don't come in cereal boxes. They are the product of a careful review process, and that process starts with a clerk like me. I think those guys wrote and typed those documents themselves.

The problem was that, rather than soldiers, they were merely over-aged clerks in uniform, who'd joined after the Korean War was done, on the assumption that they could serve a safe 20 years, earn a pension and retire, unscathed. The Peacetime Army often produced such outcomes.

I couldn't come right out and accuse several senior career NCOs of fraud. They'd cooked up this scheme during some drunken evening at their little club, but overlooked that one essential detail. A Purple Heart could only be initiated by a medical officer, who would give the wounded soldier a form to turn into his Awards and Decorations NCO. No form, no Purple Heart.

Captain Thomas lifted the entire stack, turned ninety degrees in his swivel chair and dumped them all into the gray steel wastebasket next to his desk.

"Hamit. If any of those 'gentlemen' wish to discuss these awards with me and why they will not be forwarded, I will be happy to enlighten them!" He then stormed out of the room, cursing a blue streak.

Captain Thomas

Word got around, and no one asked after them further, but my NCOIC became hypercritical and looked for new and interesting ways to chew me out. I found this curious because he was ASA, not Regular Army. But he was also a serious drunk, and those guys were his bosom buddies.

This I learned from his Assistant NCOIC who would sometimes look at him and smirk in a certain way familiar to me from my days in the theatre. So much so that I wondered if they had something going on on the side. The NCOIC was a married man and could hardly seek relief in Ben Xe Moi. Adultery is a crime under the UCMJ. But he might have been involved in another UCMJ offense: sodomy. The younger man had more than a hint of lavender in his manner and speech. Had you asked the NCOIC if he was punking the younger man, he would have denied it vehemently. He was no queer. Not while he was sober.

He had not participated in the phoney award scheme. He had standards. But he had not talked them out of it either, and he was the Acting First Sergeant.

Having spent six years in theatre, and known many gays, in and out of the closet, I was inclined to live and let live, but his posturing and rough manners told me something else about him. He was scared shitless. He had served 17 years in the Army and never been in a combat zone before. And he was supposed to lead us if the balloon went up again.

This had nothing to do with his sexuality. It had to do with his courage. The more they yell and posture, the less they have of it. Fear is contagious, which is why you don' t want them in leadership positions. If it seems like I'm being overly judgmental, remember that I grew up with this stuff. It formed my character.

When that same Captain worried aloud about my morals because I spent my days off in Ben Xe Moi, I told him not to bother. "I don't have any." His guidance was unwelcome. And sex, lots of it, with multiple girls, was my refuge from the craziness and stress. I didn't drink much because if there was another attack some of us would have to be sober enough to fight.

The most bizarre aspect of the new post was the maid service. Not what you expect in a combat zone. And it was mandatory! The Army had no facilities there that allowed you to wash your uniforms and underwear yourself. You were assigned a mama-san, a middle-aged Vietnamese woman, who made your bunk, cleaned and ironed your clothes, and shined the leather on your jungle boots. She was paid the equivalent of four dollars a week in piasters. This was enough for her to feed her family, and then some, since she had several clients. This was part of a plan to boost the local economy and buy Vietnamese hearts and minds. You had to take care of your own locker, which was always locked, because you stored your money, liquor, and M16 there along with your other gear.

I had a hard time eating anything at first. The food exceeded Army standards and was served hot. Lots of local vegetables were also provided. It was delicious. There was nothing wrong with any of it. The oppressive heat killed my appetite, and I would leave Vietnam weighing only 165lbs, ten less than I'd weighed in high school, and 20 less than when I'd enlisted. My body produced massive amounts of ketones during the change, which meant that I developed a severe and unpleasant body odor which showering did not diminish. Not that the showers were working all that well. They were usually a trickle of cold water when they worked at all. So I came in for some negative commentary. Someone left Dial soap on my bunk. Eventually my body acclimated and the problem went away.

But it made the prospect of a steam bath on my day off very attractive. Especially when administered by a beautiful Vietnamese girl. There was a quasi-legal area of steam bath bars and other businesses close to the base called Ben Xe Moi. There were three dozen of them, and all were also brothels.

The price of a piece of ass was standard at 200 piasters. That was $1.89 US at the then-exchange rate. Beer cost about half of that, and the steam bath was 50 piasters if they bothered to charge. Usually a pretty good massage was included as part of the package, all of which was delivered in between 20 and 40 minutes. There was another bonus. The girls were among the most beautiful in the world, and usually not bored professional whores, but enthusiastic amateurs who seemed to like the work. They would actually kiss you, which is not the custom in most places. If you treated them well, they reciprocated. And if you actually knew what you were doing in bed, word got around. Some of them were also students at the local university, and spoke good English. ASA guys tended to avoid those, because we assumed they were VC spies. Besides, we weren't there to talk.

Undoubtedly some were VC spies, because VC intelligence would have been remiss had they not exploited such an obvious channel. Intelligence is the second-oldest profession, and one that sometime walks hand-in-hand with that one, but most of them were just trying to feed their families. All of them had sad stories to tell, some of which might have been true.

I got one day off in 14. I spent them all in Ben Xe Moi. I was already something of a sex addict. Here, I could satisfy my lust three or four times in an afternoon, with as many different girls. And be teased for being a 'butterfly' for flitting from flower to flower. The one thing that you could not do is fall in love with one. There were some Regular Army guys who did, and married their favorite squeeze, and others who romanced 'respectable' girls who worked as VGS secretaries in our offices, and took them home, but not ASA. Fuck them if you will, but don't form a relationship, was the unofficial word.

In time, I would spend my Saturday afternoons with one particular young lady. She spoke only rudimentary English, but after our first sexual encounter asked, "You want to go again?" and twenty minutes became four hours. We found ourselves perfectly matched, which is a rare thing indeed, and not to be despised nor wasted. No extra charge then or later, much to the irritation of her employer. Somehow we moved beyond mere sex into making love, and our time together was a refuge for both of us, from the war and the world outside. I would bring her two packs of her favorite cigarettes, and tip generously, but she was losing money on the deal. I took that as a compliment.

At night the men either watched the movie, which came two nights a week, or went to the Enlisted/NCO club, a huge barn-like structure with cheap beer and occasional entertainment.

Usually this was a local Vietnamese band, either rock or country-western, or an American one on tour with a USO program. Sometimes the act would be a girl singer, and she would be the subject of intense attention, because she was a 'roundeye' who looked like the girls back in the States. I recall one, dressed in a form-fitting black leotard and tights, who sang folk songs and had a hunted look in her eyes, because the men thought she was a hippie. I suspect that the local club NCO had other things in mind for her than singing, when it came to 'entertaining'.

The big acts did not get this far south. We would have loved to have seen Bob Hope and the pretty girl entertainers he brought with him. We never did.

CHAPTER 16

We had a change of command ceremony, trading one over-aged Army Aviation major for another. All of the commissioned officers were on six-month rotations for 'career development'. They would spend just enough time to learn a job, get somewhat proficient, and then move someplace else at that company or another. It was called 'getting your ticket punched'; it played hell with unit cohesion when you factored in the constant comings and goings produced by the standard one-year tour for all of us.

During the change-of-command ceremony, the outgoing major, a kindly older man everyone loved, was honored by a fly-over of four of our U-6 Beavers flying wingtip to wingtip. The pilots were showing off, of course, because this was a stupid and dangerous stunt that put two-thirds of our aircraft at risk. My NCOIC cursed them under his breath when he saw the formation float by at close to stalling speed. The flight plan filed did not revealed this sentimental gesture. But they pulled it off, and the new commanding officer shrugged it off, and told them to never do it again.

By May, the war in our area had settled down. The action was out on the Mekong River, and we were graced with the first Army Huey Cobra company assigned to Vietnam. This was a fearsome machine, and one or two of them flew over the airfield constantly at night to ward off a surprise attack by the VC. We also had a Special Forces support element on the airfield, and those guys liked to keep their field skills fresh. They would take little strolls at night beyond the wire. One night they discovered an entire VC regiment, perhaps as many as three thousand men, setting up for an attack on the airfield. Fading back, they alerted the Tactical Operations Center to the location, and got the hell out of there. Two Cobras were overhead. One went left and the other right, and each made a single pass over the

VC formation, all guns and rockets firing. There were some secondary explosions. The Cobras came back in to reload, replaced by two more flying over the field. There was no attack. The next day scouts found that the VC had removed most, but not all, of their dead and wounded, so there was no juicy body count to forward to higher headquarters, which is probably why it escaped notice in the media.

The Cobra automatic cannon and machine guns put a bullet or shell in every foot on a hundred meter front, and shredded them. That unit was wiped out in a minute and a half. They sent out bulldozers to bury the dead that remained. It was said that very few of the bodies were whole.

We still had mortar attacks, but they became just another form of bad weather. Proof that humans can adapt to anything. I settled into a routine of office work and courier duty. It was a little boring. I was getting an extra sixty-five dollars a month combat pay, and felt a little guilty, because the guys in front-line combat units got no more. But that was the reality of the war. We were part of the other ninety percent. The 'clerks and jerks' who operated the very long logistics train required to get them into battle. No one ever writes about that, because it's not very interesting. You don't get Purple Hearts for paper cuts.

The reality was that ASA in Vietnam saved thousands of lives with timely intelligence, and would have saved many more times that, had certain commanders taken us more seriously. Because we could not discuss our 'sources and methods', and sometimes did not immediately respond to their tactical requests, they didn't trust our product. NSA wanted everything stove-piped to them for further analysis, afraid to guess wrong. There were no computers, and the radios still had vacuum tubes, and were portable only in the sense that one man could, with difficulty, carry one.

CHAPTER 17

The company never came together as a unit while I was there. Part of the reason was the one-year tour, and the constant rotation of personnel in and out. By the time you got really good at your job, you were on your way home. But we knew we were making a difference, even if we were the most coddled Army in history.

People had refrigerators in their bunk area. They sold them at the PX. Not surprisingly, some of them ended up in Vietnamese homes and businesses. There was a lively Black Market.

We were paid in MPCs (Military Payment Certificates) printed by the US Treasury. We could only spend these on base and at the PX, and were required to convert them to piasters, the local currency, to spend off base. The conversion rate was less than a hundred to the dollar. Five hundred piasters would feed a local family for a week. So while we considered ourselves miserably underpaid, the Vietnamese thought of us as spoiled rich kids out on a deadly lark. The Black Market in currency and our 'luxury' goods ruined any chance they had at a rational economy.

Possession of US currency was a court-martial offense. My mother included a five dollar bill in the card for my birthday, and I had to hurt her feelings by immediately mailing it back to her. I could not afford to keep it. She never understood this.

MPCs accumulated in Ben Xe Moi because they brought a premium price there. Those were used to buy PX goods, which were sold on the local economy at huge mark-ups. The US authorities would occasionally change the game by declaring old MPCs invalid, and issuing new ones. The base would be locked down, with no one in or out, and Vietnamese entrepreneurs would gather outside the fence carrying large brown grocery sacks stuffed with MPCs begging someone to take them and exchange them for them. The MPs would arrest anyone who tried, and bags of money thrown over the fence lay there untouched. Vietnamese, some ruined financially now, would stand and weep, calling us 'Numbah Ten'; the worst slang term available.

GIs who turned in excessive amounts of MPCs were also interrogated as potential black marketeers. One of our guys had to collect affidavits from several of us, explaining that he was simply the best poker player in the unit, and had come by the three thousand dollars in his locker honestly.

CHAPTER 18

The officers always got the weekly movies first, and they would come out to the base the next day on the regular courier run, along with the outgoing mail. This was an MP function, because there would also be locked mail bags full of Top Secret going to the Ops Center to be prepared to be flown to Saigon the next day.

One afternoon, the movies didn't make the trip, and it was the two television series. No one wanted to miss "**Star Trek**" or "**Mission: Impossible**" – and they were due to be handed off to another unit the day after. So I was told to go get them, curfew or no curfew.

Curfew was at 1800 (6 p.m.) just as dark was descending, and the Viet Cong began to roam around looking for mischief. Obviously, I could not go alone. So I assembled a fire team of three other guys to ride with me. One of them brought one of the stolen M60 .30 caliber machine guns we kept in the motor pool, and a couple of belts of ammunition. The others brought their fully charged M16s, and I had mine and my .45 pistol. We were pumped!

This was as close to combat as we were going to get, short of another attack on the airfield, and even then we would be the last called. Too valuable to risk. The sergeant who sent us on this extraordinary night time mission got his ass chewed the next day.

One of the other three was my good buddy, Bill, our 'John Wayne' MP. He was a skinny 19-year-old kid from Tennessee, and enthusiastic about the war. He appeared for guard mount not just with his M16 and .45, but also with an M2 .30 caliber carbine he'd acquired on the black market, a 12-gauge short-barreled shotgun, some hand grenades, and at least three knives sharp enough to shave with. Even the other MPs thought this excessive, and made fun of him, but he was eager and spoiling for a fight, so everyone left him alone. Most of us had an extra weapon of some kind. Mine was a wicked-looking switchblade knife which I usually used to open mail, causing more than one raised eyebrow.

Bill and I were close, I think, because we were both outliers. Despite the difference in age and education, we could talk, and we respected each other. He was smarter than he knew.

We drove very fast to the officers' hotel, speed limit be damned. It was very quiet, and no one seemed to notice a lone American jeep on the road. I had to present a written pass at the gate, and the surprised sentry on duty muttered, "It's your funeral," and waved us on.

When we arrived, the MP on duty was astounded, but let Bill and me go upstairs to where the officers were having their usual drunken party. I saw my Awards and Decorations OIC sitting to one side, trying to read. He looked miserable and probably was. There is nothing more disheartening than being the only sober person in a roomful of drunks. And the drunks included the Company Commander, the second we would have of three the year I was there. Everyone liked and respected this man. He was one of the good guys, and knew how to lead, but seeing him acting like W.C. Fields – absolutely hammered – well, it was depressing.

"Hamit," he asked, "What are you doing here?"

"You have our movie, sir."

He stared owlishly around. "We do?"

"Yes, sir. The men are waiting, sir." I was afraid, deathly afraid, that he was going to invite Bill and me to join the party. He was in a very good mood.

"Well, we can't have the men kept waiting," he slurred.

I spotted the two film containers in a corner, motioned to Bill to grab them, and we got out of there.

Going down the stairs, Bill chuckled. "They really tie one on, don't they?"

He was an MP, and sometimes worked the guard posts there, so I figured he knew how often it happened. Too often, apparently. No man is a hero to his valet – or his security guard.

"Don't say anything. Okay?"

"Never. The Sarge would have my ass."

It explained why the officers left the men at the airfield to the senior NCOs in the unit, but they were no better – in fact, worse. One had a habit of getting drunk and then inspecting our positions on the perimeter at night, yelling loudly at those on guard to stay alert. Had he ever pulled that detail himself, he would have known that the first time you heard a VC sapper cut the barbed wire, it rang like a bell, and you would never be able to sleep on guard duty again, even with your trusted buddy keeping watch. He was finally ordered to cease and desist with these nighttime 'inspections' which did little but reveal the exact location of our bunkers to the enemy. The CO was told, quietly and unofficially, that his next trip would be his last, likely due to a tragic misfire of a startled guard's M16.

We got back to the airfield without incident, to Bill's disappointment. He really was still a boy. The reality of Vietnam had not yet penetrated his mind.

The guys enjoyed the two television shows, and the sergeant who'd sent me, not realizing the ration of shit he would get the next day from the Executive Officer, thanked me for making sure they were not missed. It was little things like that that kept morale up, he said.

CHAPTER 19

Except for the encounter with my CO, a man I liked, I'd found the whole thing exhilarating. It was almost as exciting as a courier flight. I normally flew these in the Admin bird with a Warrant Officer, who was also from Iowa City, but a native, rather than associated with the University.

Just the two of us, normally, and we would fly up to Saigon and stay overnight. That meant I had a few hours to see the city, which was, as I wrote my parents, the safest in the world at that time. It had to be. There was an armed guard on every corner, and the whole place went on curfew at 1800, just like Can Tho.

One time I took a cyclo over to Tu Do Street, famous for its bars and brothels. An interested tourist. The trip there was surreal. A cyclo is a three-wheeled bicycle with a basket with seating for two in the front. The driver peddles you to your destination. In downtown Saigon, you had vehicles ranging in size up to five ton trucks looming at you from every direction, with horns blowing and tires screeching. This was as close to dying as I ever came in Vietnam, and when I got there, I naturally had to avail myself of some female companionship. The poor little thing went into terrified giggles when I pulled the .45 from my shoulder holster and put it under the pillow, but I was not about to risk losing it to a thief.

I got back to my temporary quarters, and joined the pilot for a beer on the rooftop garden of the Cholon hotel that served as temporary quarters for visitors. The fraternization rules were a little looser for Warrants, and we were MI, so the Regular Army MPs left us alone. The pilot was simply that, a guy who like to fly, and who had joined the Army to improve his skills. The war didn't worry him much, and the operational missions were boring, floating through the sky on a heading as the E-5 in the back listened for any kind of enemy signal. It was more interesting near the Cambodian Border, where North Vietnamese Army, and even Chinese signals, could be acquired. Then there might be some real flying, even 'over the line', as they tried to record the traffic and get a better location fix.

Here I will turn the narrative over to John Reid again:

It is, as mentioned, a dipole antenna. The aircraft had one on each wing. It was used to locate enemy transmitters. When an operator found a target on his radio the audio was transferred to the pilot. The pilot would then maneuver the aircraft in such a manner as to get a null in the audio. This was done by pointing the aircraft directly at the target transmitter, so that each antenna was at an equal distance ftom the transmitter. When the null was reached, the pilot would call out the bearing the aircraft was headed, and the copilot would plot this on his Plot Board. This was called a bearing. The pilot would fly at a right angle to the heading and shortly turn the aircraft toward the transmitter again, and again when the null was reached he would call the bearing. This reading was called a Cut. This routine was followed again, and the third bearing made a Fix which was where the transmitter was located. Each target had a priority, and the highest priority would generate an automatic redirect of any strike aircraft within 100 miles to attack that target. This included Arc Light Strikes which were B52 bombers with heavy loads.

Further note: The missions started at between 2,000 and 2,700 feet above ground level (AGL). The second and third passes were done at lower altitudes, because the pilots dived to gain speed and get a more immediate fix on the enemy transmitter's location.

Recognition

I only received two messages from operations concerning the results of any of the missions that I had participated in. The messages came from higher headquarters, and as a guess by me, MACV, Military Advisory Command Vietnam. Who else would have all the information? The messages only identified the mission number and date and results, and did not mention names. It required that the mission and names be looked up locally in our records to see who the three persons were on a particular mission. I thought that it was peculiar at the time. I just looked at it as my job. Finding the target transmitter is what we did. What someone else did with the information was not my concern. It didn't take me long to accept my share of the responsibility of what happened to those targets. After all, it was us or them. There was no middle ground that I had any input about. That was way above my pay grade.

U6A Beaver, ARDF

Because I flew on so many missions, I could not in my own mind read the message and say, "Oh yeah, I remember that." All I could do was accept the information in the message. Someone had actually gone to the location and made an assessment of enemy casualties. Both these messages were after the TET Offensive, 2/1/1968 and before I left, 6/15/1969.

The first message advised of a job well done. 197 body count. I remember looking at that. As I remember that, I was stunned. I thought, good, that's why I'm here. I wanted to make a difference, and I did. After all, the target was Northwest of Vinh Long, South of the Parrot's Beak, 'A description of the Cambodian/ Vietnamese border to the West of Saigon'. The area didn't have many residents. From the air, it just looked like miles and miles of elephant grass with a few huts scattered around. It was difficult to get good bearings there based on visual ground references. We tended to hang around a good reference point on the ground so we could get a good bearing if a target started transmitting. With a body count of 197 people in one place, they were up to no good. I didn't think twice about it.

The second message also advised of a job well done. 3 body count. Again I remember looking at the message and being a little bit stunned, but for a different reason. I didn't recognize the mission. The location was across the river from Vinh Long and a few miles North. My first thought was, how do we know these were bad guys? Maybe it was just some locals having lunch in the wrong place. Bad timing. Then it dawned on me, they were transmitting. I always had call signs with my fixes. They were bad guys. They were up to no good.

The fate of my targets I helped locate never got in the way of finding them. I had no illusions on what they were trying to do to South Vietnamese, American, Australian, or South Korean forces. There was no middle ground. If our flight was shot down, I expected no quarter, and neither did they. It was us or them. It's just the way it was.

The public recognition for mission personnel were Air Medals. The top secret work was recognized by higher authority. They knew that the work was dangerous, and always over the enemy, within small arms range. Every mission was considered combat assault

On the next courier run, we had a passenger. There was a big battle in progress next to the Cambodian Border, and a temporary airbase was located at Muc Hue. This was a bunch of metal plates dropped on top river sand on top of rice paddies, same as Can Tho.

On the way there, he said casually, "Take the controls. I have to look at the map."

"What!"

"Just put your feet on the pedals, take the stick, keep it level and follow the river. I want to make sure we don't end up in Cambodia."

Gingerly, I did as he asked. I later found out that this was a trick he played on all the couriers. Bill, my MP buddy, grabbed the stick so hard that the plane surged upward and then pushed it so hard the other way that the warrant officer had to grab the control back.

"Ever hear of a thing called 'stalling'?" he had asked, the joke having gone sour because his victim had come close to making the U-6 fall out the sky, which would have killed them both.

But I got 20 minutes of flying time in, before he took the controls back. It was not relaxing, and I decided right there that being a pilot was not for me. I didn't have the requisite 20-20 eyesight, anyway.

"Good job. Call the tower." In the back, our passenger resumed breathing. He was a fire chief going to Muc Hue to install a shiny new fire engine. This temporary base was that new.

I took the radio and gave our 'Lonely Ringer Six' call sign, and was given an altitude and a heading that put us between two C-130s preparing to land. We could see the airfield, but not the tower. The C-130 is a magnitude of order larger than the U-6, and we had to trail way behind the first one to avoid having the backwash from its engines flip us over.

"Where the hell is the tower?" the pilot asked several times, looking around as we taxied to a place where we could drop our passenger. C-130s full of supplies were landing at the rate of one a minute, unloading and taking off again. Can Tho Army Airfield, which had a big, well-manned control tower, was never this busy. I spotted a large wooden high chair at the side of the runway with one guy holding a radio mike and another using a big pair of binoculars to scan the sky for incoming aircraft.

"Sir?" I pointed at this as we rolled slowly by to get off the runway. "Air Traffic Control, Muc Hue."

His jaw dropped. "Jesus, Mary and Joseph," he muttered; "Let's get the fuck out of here!"

We dropped our passenger off, and did just that. He joked about going back to Can Tho with us, but we'd already stretched a regulation just by having him along. It was a security violation for non-ASA personnel to be anywhere near those locked mailbags. And our next stop was Saigon.

It was a bright clear day, but the pilot couldn't go directly there. He had to fly around an area where artillery fire was being delivered in support of an operation. It took extra time and fuel, and worried him because we didn't have that much. He cut his airspeed to use less. As we floated leisurely over the Delta, we saw a huge red gash in the jungle below. About a mile wide rectangle a mile and a half long. Total destruction. Nothing could have survived.

"Rolling Thunder," he commented, "B-52 strike the other night."

I remembered it well. The entire ground at Can Tho shook like a minor earthquake because the Delta is one huge mud pudding. We saw the flashes of the bombs hitting on the horizon, and heard the muffled explosions that gave 'Rolling Thunder' its name.

"VC regiment," he added, "Off of our intel." Then he looked away, realizing that he'd said too much. I wasn't supposed to know that, and neither was he. That was one of the problems with the constant boozing. People talked and compartmentalization broke down.

There was 'no ASA in Vietnam'. No one was ever supposed to figure out what 'Radio Research' meant, and it meant different things to different units with different pieces of the mission pie. Our gloss on it was that we were engaged in the war's biggest littering campaign by dumping propaganda leaflets on the enemy. We'd shown some left-over ones to the fire chief and let him take one as a souvenir.

Sometimes the wind would shift and a bundle would float all over the airfield, which would generate a written rebuke from the Airfield Commander's office. At the same time, we got a rather sarcastic note from Psychological Operations, reminding us to untie the bundles before throwing them out of the aircraft. One had been found, intact, in a tree. You couldn't win.

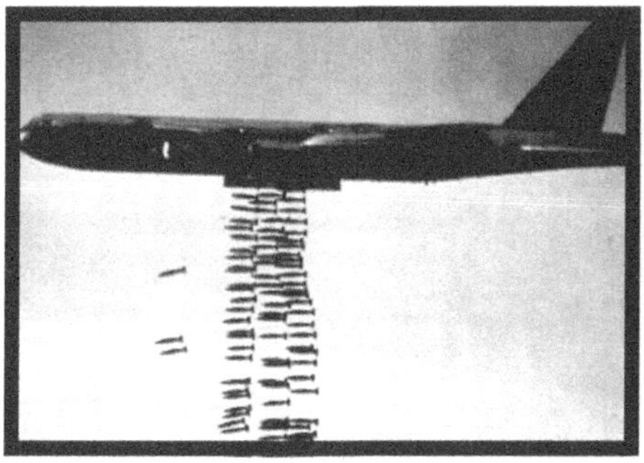

Arc Light bombing damage, photo by Col. Pete Bradley, 146th Aviation Company (RR)

USAF photo of bombs falling from a B-52

The leaflets were part of our cover. We simply refused to discuss what else we might be up to. Classified, don't you know? But sometimes the cover slipped anyway.

CHAPTER 20

One day, on my usual mail run to the post office next to the control tower, I ran into The Preacher, my buddy from Basic Training.

He'd been to OCS and was now a Second Lieutenant. I was still a PFC (E-3). If that was not awkward enough, he recalled all too well where I'd been sent after Basic.

"Hamit," he said, his amazement at seeing me plain; "What are you doing here? Weren't you going to some top-secret outfit like ASA?"

There was no ASA in Vietnam. The Army said so. Now what?

I looked down at my dirty jungle boots and muttered, "That didn't work out, sir."

"Really?" He had big black-framed GI glasses, and was wearing the patch of a local transportation outfit. Not combat arms, but not MI, either.

"I had a little trouble... ," I let my voice trail off, and then added, "I took Vietnam to get out of it."

"Oh, sorry to hear that. Well, now that you're here, we'll have to get together some time."

I could tell that he still had a bit of the preacher left in him, despite now being a twice-trained killer, and was hatching some scheme to 'bring me to the Lord.' I wanted none of it, and couldn't have him hanging around our barracks. So I went Army on him.

"I'm sorry, sir, but with all due respect, that's a really bad idea. You're an officer now, and there are rules about fraternization. My CO is very strict about that. West Point. Besides, I spend my off-duty time chasing whores, and you're a married man."

He stared at me, a bit hurt. "Not any more," he muttered, "She divorced me when I got orders for here."

I could see and feel his pain, but had to stay firm. "Sorry to hear that, sir." I saluted him, making my point. He returned the salute and we went our separate ways, never to meet again.

I reported the encounter to my First Sergeant, of course. By this time Ezio Tozzi had taken up that post. One of the first things he said in his initial speech at one of our rare formations was that we were to come to him with any problems. The Preacher was definitely a problem. So was the junior enlisted man I encountered in the mess hall one day.

He spotted my circular plastic-encased 156th Aviation Company patch, and took a step back. I suppose I must have reacted with mild surprise in some way.

"I know who you guys really are," he said, very nervous.

"Excuse me?" The other guys in the line weren't paying any attention – yet. So I just looked at him, pretending to be a little bored. "Who are we?"

"ASA. Army Security Agency." He trembled as if he were ready to bolt and run.

"Who? What's that?"

"You're not ASA?"

I shrugged. "Never heard of it."

"Bullshit! You guys are ASA!" His voice was rising now. I took note of his name and unit patch. He was in the 235th Aerial Weapons Company, the Huey Cobra outfit. I kept my best poker-playing face on. He was obviously afraid of me now, but the last thing I could do was to exploit that fear. No one else in the line had heard him, and the last thing I wanted was an argument in a big room full of Vietnamese civilian workers. Another part of the scheme to buy 'hearts and minds' by providing above-market wages to the locals. The Regular Army cooks were all supervisors now, and when we got one in from ASA because the company TO&E said we had to have one, he was reassigned to the motor pool or the supply room to learn a new trade.

So I turned away and ignored him, while he trembled with indignation at having to suffer the company of a spy. After I finished lunch, I made sure to stop by our Orderly Room and have a quiet chat with Ezio Tozzi about this young soldier. He listened carefully, and said he would talk to the 235th's First Sergeant about loose talk in the chow line. The ironic part of this was that communications security was not part of the 156th's mission. We did ARDF and intercepts of enemy forces, and that was quite enough.

A word about Ezio Tozzi. He was one of the good guys, and a great First Sergeant. He was a traffic analyst by trade who'd been required to take a Top Kick position because someone had to do it, and it had to be someone from ASA, because Regular Army people, cleared or not, were not permitted to know the things about ASA that he already knew. He was a very kind, soft-spoken man, who'd been in the Army a long time, and he didn't drink. He'd politely refused to join in the nightly bacchanal with our other senior NCOs. He wrote poetry in his spare time.

First Sergeant Ezio Tozzi

"They had every kind of liquor," he wrote me in a letter near the end of his life, when we reconnected and I asked him for help with this memoir.

And they got hammered every night. For the most part, they did it in private, which was a good thing. I recall coming out of the EM club one night into a pouring rain and finding our Motor Pool NCOIC, an E-6, passed out, and laying face down in a puddle with the water rising. I didn't much care for him – Regular Army, and another hillbilly who despised 'collitch boys.' I tried to wake him, to no avail, and was not about to try and lift his soaked, slippery body, which weighed more than mine, and get him back to his bunk. So I settled for turning him over so he was face up with the rain still pouring on to his face. Either he would wake up or catch pneumonia, but he wouldn't drown. Turning him over may have involved a couple of stiff kicks with the toe of my jungle boot. I don't really remember.

Somehow, he had a broken rib the next day, but he fell down a couple of times getting back to the barracks, I was told. I didn't hang around. I think it was at this point I really began to despise drunks. Not people who occasionally drink too much, but people who make it a practice to get so hammered that they are unable to function. Our pilots and operators never missed a mission because they were too hung over to fly. Our MP platoon was always short-staffed and overworked, and observed a one-beer limit.

These Regular Army imports were frequent visitors to sick call for headaches. Must have been all that aircraft engine noise.

Had the troops at the 235th known it, our mission was even more critical than theirs. They delivered fire support. We found out where to put it, along with a hundred other pieces of vital intelligence, day in and day out. But the last thing we wanted to do was advertise that fact.

To maintain cover, we pulled our full and complete share of base chores. We manned our assigned part of the perimeter at night, we provided trucks, drivers and guards for supply convoys, and would have pulled KP as well, had not all those jobs been reserved for local civilians. This created another kind of security problem. Vietnamese wandered the base freely, and some were VC scouts pacing distances for targets, or coming into the barracks trying to sell things and identify key personnel.

The first month I was there, I was in the barracks alone, having pulled guard duty the night before, and getting ready to report to my office for the afternoon, when this young man, a slender handsome Vietnamese youth in the uniform of an ARVN 2nd Lieutenant, came in and engaged me in conversation. He wanted to sell me an ARVN uniform, saying that he needed money to feed his family. That might have been true. ARVN troops sometimes went weeks or months between paychecks because of a combination of corruption and incompetence, and everyone was in the Black Market. But it didn't sit right with me. Not only was it neither my problem nor my responsibility, the uniform was provided to his government as part of our military aid. ARVN troops looked different from ours only because they didn't yet wear the new jungle fatigues. A point that the Viet Cong exploited for propaganda purposes.

And he was making noises about becoming friends. That didn't pass the smell test either.

"Sir. You are an officer," I said, with alarm bells going off in my head; "In our army, officers and enlisted men are not allowed to be friends. It's considered prejudicial to good order and discipline."

That wasn't the reality, of course, but he really worried me. Either he was looking for a Black Market contact, in which case I was being set up for blackmail, or he was an enemy agent. Of course, it might have been perfectly innocent, but he was too friendly, like a car salesman going for the close, or he wasn't really an officer. I was a Private First Class, so there was little I could do about my suspicions. Ezio Tozzi had not yet arrived and I had already learned not to bother his predecessor with 'trivia'. I excused myself and escorted him out of the barracks, saluting him when we got outside. From his embarrassed reaction and poor rendition of his own salute, I got the idea that he wasn't really ARVN. He didn't have a 'cover' (hat) either . The ARVN usually weren't much in a fight, but boy, did they know how to salute! He never came back, so this is one of those mysteries I will never solve.

CHAPTER 21

In the interests of maintaining our cover, we eventually had to accept a VGS (Vietnamese General Schedule) civilian secretary in the Flight Operations office. To refuse her on security grounds would have been like erecting a big red sign that said: 'Target!' or: 'Here we are. Spooks are Us!'

She was a charming young lady whose English was very limited, and was said to be the daughter of a high-ranking Vietnamese Air Force officer. Now that we had her, what could we do with her? The most secure path would have been to give her a book of crossword puzzles and call that training. We settled for having her type the recommendations for the Air Medal that accumulated weekly. These had to be rendered perfectly to satisfy higher headquarters, and only had the flight hours logged by the pilot or operator, day by day. No other details, such as mission or type of aircraft or other crew members, had to be provided. If you put in the time over a combat zone, the time accumulated, and you got another Air Medal for every 24_hours of flight.

To my disappointment, the hours I logged as a courier didn't count toward an Air Medal. It was not 'combat time' but 'admin'. It was not so much the trinket itself as the points it generated for promotion.

Miss Toi, our VGS secretary, would report for duty. I would hand her a stack of forms with the name and numbers written in, and then she would replicate them on a typed form. Fortunately, she was as bad a typist as I was, and many had to be redone. No one talked Flight Operations in the Flight Operations office when she was around. Another reason I never really learned that job. I never had to try to find additional work for her, and, for security reasons, there was none I was comfortable with giving her, unless it was requests for USAFI (United States Armed

Forces Institute) correspondence courses. Those were few and far between.

These were free college-level courses for transferable credit that any officer or enlisted man could request, and work on at his own pace. As Education and Training NCO, part of my duty was to sign up a certain number of guys every month to take one of these. Not surprisingly, no one was much interested, but after I explained that they were not obligated to actually complete the courses, and it would look good when they went for promotion, a few of my buddies helped me out and got free textbooks in the mail. I signed up for 'College Algebra' myself, but was as hopeless about math as I had been in high school. This program was another illustration of the utility of certain stateside requirements in a combat zone. But we had to play the game.

One blistering hot day in August, I got a day on base chores, second only to working the shit burners for misery: the garbage truck. The base used a civilian contractor with, of course, a Vietnamese crew. The truck had a route that included not just trash, but mess hall trash. Due to Army Regulations on food storage, a lot of perfectly good food went in the garbage along with the considerable pile of leftovers. (Food was prepared in stateside quantities, but as previously mentioned, no one had much appetite because of the heat.) Number ten cans of food were dumped , and to prevent the VC from appropriating them at the dump, and using them as a source of supply, the cans were supposed to be opened and the contents dumped. An armed guard was supposed to keep any useable food from being diverted. Usually, you had to haul your M16 along, but I had that .45 in a shoulder holster. I took that instead.

When the truck appeared at the gate, the smell of rotting food nearly knocked my off my feet. There were six middle-aged Vietnamese men hanging onto the sides, with the crew boss next to the driver. I was shown to the place of honor on top of the cab, where I could really see – and smell – what was going on.

We rolled slowly from building to building, with the crew dismounting and tossing everything that couldn't be salvaged into a big bin at the rear of the truck. Then the compactor would sweep it all into the belly of the beast. I took an occasional swig from the canteen on my web belt, and watched the operation carefully. The men on the crew sometimes looked at me, a little worried. Especially if they took some salvageable item such as a broken office chair and set it aside. I admired their enterprise. The Army no longer wanted it, and they were all natural mechanics and handymen, so it was a fair catch. Somehow you couldn't see that ending up in a VC bunker out in the field, and you didn't think about the extensive urban operations that were so useful to them during the Tet Offensive. More worrisome was the number ten cans of discarded food, where a cook would come out, use a pick-ax to puncture each one and throw them into the truck – like the VC had never heard of candle wax to reseal them with.

But I didn't want to hassle with the cooks over something so theoretical. When we got to the 235th's mess hall, the pleasant if odiferous day became suddenly less so. As the crew was pouring the wet garbage into the back, not one, but two brown rectangular items caught my eye, to quickly disappear under more of the leftovers from lunch.

"Hold it!" I said loudly, and got to my feet. I pointed at them and they all looked up at me, alarmed, and stopped dumping trash on top of the load. I said a few crude Vietnamese phrases, and indicated by pointing that they were to stand next to the mess hall wall. When they hesitated, I touched the .45 in its shoulder holster, and they scrambled to comply. Then I took a deep breath and started digging though the wet garbage, looking for the two Army manuals I was pretty sure I'd seen buried.

It took a few minutes, and a couple of them wandered back, peeking over my shoulders. Finally, I found them. They were two slightly damp but very readable technical manuals for the Huey Cobra. One labeled 'SECRET' and the other 'CONFIDENTIAL'. An intelligence bonanza for the VC. I tucked them under my arm, and walked to the 235th's Orderly Room. I was wet to the elbows with fermented garbage.

The question in my mind was how much of a fuss to make about this. Obviously, at least one of the garbage crew was a VC scout or spy. Probably more than one. Maybe all of them. I'd moved too quickly away from the truck for them to react. I pointed them back to the wall and shifted the .45 slightly without drawing it, so as not to make too much of the discovery and keep them intimidated.

Here was my dilemma: I could have all of them in the POW camp within minutes by calling the MPs. Who, then, would finish picking up the garbage? It would pile up, and I would become known as the eager-beaver from the 156th, who'd created a sanitation problem for the entire base. That would call unwanted attention to my unit, and I was only on this lousy detail to prevent that.

I walked into the Orderly Room. The 235th's First Sergeant smelled me before he saw me, and all of his clerks turned and stared as well. He was wearing a Class A Army Green uniform with bloused trousers and well-shined black combat boots. A poser. That uniform was only permitted Special Forces personnel outside of the combat zone. He had no SF patch nor Combat Infantry Badge to justify this self-glorification, and his ribbons were nothing to write home about, either.

"Who the fuck are you, and what are you doing in my nice clean orderly room?" he roared. I didn't say anything, and just walked over to his desk as he stared at me, shaking with indignation. I dropped both manuals in front of him. They made a very satisfying 'splat' when they hit, face up, the SECRET one on top. He looked down and turned white. I think his entire career flashed before his eyes. I didn't care. I was hot, dirty, tired, very pissed off, and would have just as soon shot him as any of those VC garbage men outside.

"These are yours, I believe?"

He nodded, scared now.

"I'm with the 156th. Someone will be in touch."

I turned around and walked out of there, with everyone staring at me. Outside, I motioned to the garbage crew's chief.

"Okay, " I said. "Mount up. Let's finish this."

He nodded, a bit dumbfounded. No MPs? No POW camp? The crew scrambled to finish up, and we moved on. He must have thought me very lazy.

First Sergeant Tozzi gave me some quiet praise, and said it was a shame he couldn't put me up for a medal. I didn't care about a medal. I'd prevented a serious leak to the enemy about our best weapon's capabilities and vulnerabilities. I was proud and happy with myself. It was the best day's work I did in Vietnam.

CHAPTER 22

My mother sent me a homemade chocolate cake for my birthday.

I told her it was delicious and everyone enjoyed it, because I didn't want to hurt her feelings, continuing our long family tradition of social lying. I have no idea what it tasted like, because it was crawling with bugs when I unwrapped it, and it went immediately into the trash – and then was sprayed down to kill the six-legged invaders. People in the States were just clueless about what the conditions in Vietnam were like. She also sent me canned goods. Little luxury items that were also available in the base PX – and she'd been an Army wife and should have known this.

One item was these little cans of anchovies. I'd had a few jobs cooking in restaurants in Iowa City, including pizza. The smell was overpowering when that was one of the items added, and I dislike them to this day. So they piled up in my locker on the top shelf.

One night my buddy, Bill, the young MP, woke me up.

"Hey, Hamit," he said, "Have you got any food? I just got off duty and the Mess Hall is closed." I looked at my watch. It was about one AM.

"I've got anchovies," I said, trying to wake up.

"Anchovies? What are those?"

"Fish. Little fish."

"Great! I love fish. Can I have some?"

"Take them all," I said, and turned back over to go to sleep as he raided my locker.

The next morning he walked into Flight Ops and leaned over my desk.

"You sonofabitch!" he began.

I grinned at him. He looked a little sick. "Why, Bill. What's the matter?"

"I had to chip myself out of a block of salt this morning!" he replied.

"How many cans did you eat?"

"Eight."

I laughed. "One is the usual ration."

"Now you tell me!"

This was another little lesson about conditions in the field. You needed to stash some extra food, because regular rations might suddenly not be available. Most of us had a hoard of C-Ration cans. Fruit cocktail was a favorite. We were issued C-Rations every time we went on guard duty. Usually a box of twelve, which meant one apiece, and some left over to be split up. The PX stocked goodies from home, and everyone kept a can of soup handy. We might be in the tropics, but it got damn cold some nights. Poncho liners were supposed to be issued only to Special Operations troops for field duty. They were very lightweight and very warm. They went for about fifty dollars each on the local Black Market. The standard brown wool Army blanket was thin, and you were only issued one. Staying warm was a problem during the monsoon season. You could catch a cold, or even pneumonia, as I did again. Soup and whisky were the quick cure. The poncho was the long term solution.

One time I was on a detail with half a dozen other guys from the company, repairing the bunkers where we stood guard on the perimeter at night. The sandbags were rotten and had to be replaced. There were half a dozen of us with shovels and new bags, filling and stacking them, when a sudden downpour of cold rain from the South China Sea came down on us. We were all drenched in an instant, and piled back aboard the deuce-and-a-half (2 ½ ton truck), and drove back to the barracks. We had no field jackets or ponchos with us. The bunker repairs would wait for better weather.

By the time we got back, we were all wet, shivering, and even shaking from the cold, so we pulled out our cans of soup and passed a bottle of whisky around, trying to get warm, as the soup heated up on these little electric hotplates everyone had. Soon we were all warm and a little drunk, and decided to call it a day. No one came looking for us, because the rain continued pouring down until the early morning.

It was a good day, because we felt the same kind of fellowship that binds front line units together. There you have to depend on each other to survive. The military term is 'unit cohesion'. It's an important part of morale, and generates the fighting spirit needed to win battles. Most of the time it was sadly lacking at the 156th.

The military historian S. L.A. Marshall studied troops under fire during the Korean War, and concluded that it was not any particular ideology or patriotic feeling that produced acts of heroism, but the more tribal feelings of love, respect, and regard for one's buddies. On a large base like Can Tho Army Airfield, those feelings were muted at best, and the continuing culture clash between generations worked against this harmony. At the 156th, the problem was further enhanced by the division between ASA and Regular Army culture. On the surface, everyone should have been on the same page, but many of the ASA personnel were simply guys who had chosen to dodge the draft by joining a unit that they had been told was not in, and would never be sent to, Vietnam. Not all of us were so naive, but many were, and felt defrauded to find themselves there. When they complained, they were told to suck it up and reread the part of their enlistment contracts that covered 'the needs of the service'.

The 156th was a house divided against itself in many ways. The ASA were the operators and other technicians needed for the mission, plus the MPs and some of the clerks. The pilots were Regular Army aviators, or doing a ticket-punching tour for promotion. Some mechanics, motor pool, and supply guys were levied in from Regular Army units, that were all too happy to get rid of them. Perhaps this description is harsh, and overly broad, but it is the one that lingers in my memory. ASA guys tended to hang together. They could talk to each other, not about the work, but the broad range of interests they shared. Most had been to college and were the closest thing the Army had to geeks. They played a lot of poker, too. A binding experience.

Regular Army guys who'd graduated high school (or maybe not) were resentful of 'college boys' and their interests. Fear entered into the social dynamic. Not just the overaged clerks in Senior NCO dress, but lower ranks who were quick to dismiss ASA guys as 'shaky' or even 'homos.' That was divisive.

Working together on a perimeter bunker was a binding experience. Especially when the job was rained out, and everyone had to try and get warm together to keep from getting sick. It cut across the cliques that had formed.

First Sergeant Ezio Tozzi worked on this problem the whole time he was there. His was a soft-power approach, and he worked to persuade, rather than command, from above, because he knew ASA and its unique culture. He could be tough, too, and once told one chronic whiner, "If I hear you call another Newbie 'shaky' again, I will beat you like a dog." It shocked everyone, because the promise was so much against his usual nature, but he was our connection to the past, having served in the Korean War and World War II. He was a soldier who happened to be a Traffic Analyst and a First Sergeant. He led by example, which is why he didn't drink. But he was, if anything, too easy on those who did.

If it surprises you that we had liquor in our lockers, well, it was simply standard operating procedure there. The EM club didn't open until 1800, and many of us worked rotating shifts or pulled guard duty at night on a rotating roster. A bottle of Jack Daniels, tax free, sold for less than five dollars. We were short of many things in Vietnam, but the booze arrived at the PX daily. Every bar in Ben Xe Moi had PX booze and cigarettes.

I am sure the moralists in the USA would have been properly shocked at the mild debaucheries available in Ben Xe Moi, and demanded that it be closed or put off limits, but many of us relied upon it for stress relief and sanity. A 'respectable' Vietnamese girl would seldom date an American soldier, unless she was ready to leave the country and marry him.

Female companionship was readily available there. Most of the men availed themselves of these services. Many were married or engaged. Closing those bars would have put a big dent in the local economy, as well as troop morale.

Major Robert Miller

For those young ladies, whoring was usually not a matter of poor morals, but of necessity. Some were accumulating a grubstake, or looking for a foolish American to take them away from it all. And some found that they liked the work and the sweet corruptions that went with it. If not for the war, most of those young women would have fiercely protected their virtue. As the author of ***The Year of Living Dangerously*** put it, hunger is a wonderful aphrodisiac.

I encountered Catholic mothers pimping their daughters, who were too young by a year or three to work in the steam-bath bars. There was more than one way to destroy a village to save it.

Corruption was endemic. There were laws and local police, but there was also a war on, and money to be made. Capitalism. It was the American way and some senior NCOs from the base were co-owners or otherwise involved in the trade. I'm pretty sure that none of them were from ASA. We took a dim view of such ventures, and had our own cops and spies watching for them.

The Regular Army career guys dumped on us by their old units were more likely candidates, but took heed of our culture and abstained from that, at least. The tour was only a year and the work was hard. Why be clueless? But many were. Racism was still a problem, and drugs were starting to be. Sex and rock-and-roll were allowed, but never drugs.

One of our company commanders gave a sarcastic little talk one day when a newbie was caught with some marijuana he'd smuggled in from the States.

"This is really crappy blow," Major Miller said, a middle aged man trying to be 'cool'. "If you want the really good stuff you have to go to Cambodia. I'm told you can get a kilo of Cambodian Red at the source for two dollars. That's only about sixty miles west of here, so you can probably hitch a ride to the border. But leave your rifle here. We'll need that. And if you go, don't come back, because we'll be sending you to LBJ (Long Bien Jail, the big Army stockade in Da Nang)."

But the offender was not sent there. He was needed to do his job. So he got an Article 15 (Company punishment) and was busted back to Private First Class (E-3) from E-5, and lost two thirds of his pay for two months. It didn't improve his attitude. He remained clueless, and got into other trouble as the year went along. Even in ASA, the drug culture was making inroads.

The clueless, unfortunately, also included many middle-grade officers and senior NCOs arriving in country for the first time. All they wanted was the Army they knew from Stateside.

Everything clean, neat, and in good order. In Vietnam, they got a fast education in reality or died trying. In the support spaces, the higher the rank, the more likely they were to isolate themselves in air-conditioned trailers and hootches, and not experience the heat, humidity and stink that was the lot of most of us. Our pilots spent their operational time in the air, which was less uncomfortable and allowed them a certain detachment. Even a mile-long open red wound of a B-52 strike on the dark green jungle didn't seem quite real. It was a motion picture special effect in the long movie we were all forced to watch and act in.

CHAPTER 23

Stateside morals and standards were doomed to fail in the combat zone, but this reality never took hold with some people. Our next Commanding Officer was a graduate of West Point, and a very bitter man because his career progression was stalled, and he was behind most of the rest of his class. He was a fixed-wing pilot in a helicopter war, and there was no glory to be won at the 156th. He took that out on the rest of us.

He was going to bring us up to the golden standard of a stateside operation. We could tell that from the way his jungle fatigues were tailored. He was not a drunk, so the party at the officers' hotel was suddenly muted. He flew missions every day, and then spent his afternoons in his office, which was in the Flight Operations building, doing the paperwork his predecessors gave short shrift.

And the first thing he did was get rid of First Sergeant Ezio Tozzi. We never knew why, but suddenly Tozzi was gone, swapped with the 224th Aviation Battalion (RR) Headquarters Company's First Sergeant, Lowell Pyke. A few years ago, before he died, Tozzi and I reconnected by mail, and exchanged a series of letters about ASA and our mutual experience at the 156th. Here is what he told me:

He bent over to pick up and shred a cigarette butt when other men in the unit were policing the company area outside the Orderly Room. Our new commanding officer upbraided him for not calling over someone of lesser rank to do it instead, and Tozzi replied, in a voice of sweet reason, that he had always worked in ASA organizations that didn't stand on ceremony, and where everyone pitched in on such chores. Such things were his responsibility, and he would do things the ASA way, thank you, and lead by example. With all due respect.

He got orders for the transfer the next day.

"Don't I get to choose?" Tozzi asked, because ASA units in the field were usually controlled by their senior NCOs. No, he was told. It was a done deal. Pack your gear and report to the field station at Tan Son Nhut. The new Top-kick, Lowell Pyke, was told to take a harder line with the enlisted men. He mostly ignored this order as being counterproductive, and Ring Knocker (West Point culture) nonsense. But Tozzi's soda fund went away.

Tozzi kept a refrigerator full of sodas in the Orderly Room, and anyone stopping by could have one for a quarter. It was a long walk to the PX, and having sodas handy improved productivity in the offices and shops. The modest profits generated were dispensed in the form of Sunday afternoon parties with beers and steaks grilled over an improvised grill made from a cut-down oil drum. When the soda fund went away, so did the parties. These gave the men from different sections a chance to socialize and bond, not just with each other, but with guys from other units on the base. They were morale boosters. We were security-minded, but not anti-social. It was an authorized Unit Fund activity, supposedly immune from command influence. But the parties stopped. The monsoon season was beginning, so that would have also been a dampening influence.

I concentrated on doing my job, having finally been promoted to E-4. My NCOIC really had no excuse for not putting me up, except for my being someone who'd attended an institution of higher learning, and this was ASA, where some enlisted men came in with graduate degrees. He was being mean and stupid, and finally the Executive Officer intervened and ordered him to put me up. The next step was E-5, but I had to wait four months before I could be put up for that – if there was a slot open in our TO&E.

I began to think about extending my time in Vietnam another six months to grab that easy promotion. I had acclimated, and even the mortar attacks no longer worried me. I was not worried about the shell with my name on it, just the ones labeled 'to whom it may concern'. The residual fear kept you sharp and alert. Despite the problems with some of the senior NCOs, I was getting comfortable.

I was doing five jobs, if we were going by stateside standards. I was good at all of them, and would be recommended for an Army Commendation Medal by my OIC. I scared my parents when I mention the possibility of extending my time there in letters home. My father was polite, but wrote back, several times: "I wish you wouldn't."

We had an incident on the flight line. Our aircrafts' engines were partially maintained by a civilian contractor, and one day one of their technicians spilled aviation gasoline on himself, caught fire, and, panicking, ran away endangering himself and the aircraft on the line. An ASA soldier ran after him, tackled him, and rolled him in the dirt, extinguishing the fire and saving his life, and possibly a couple of aircraft as well. The technician was burned over about half of his body, and our guy was a bit singed as well.

The Army has a special medal for such events called the Soldier's Medal. As the NCO in charge of Awards and decorations, I certainly thought he was deserving. So, just doing my job, I went out and got affidavits, which I typed up and had signed by witnesses. The new company commander refused to approve the award or to forward it to higher headquarters.

When I protested, he said that too many awards were being given out. The unit had just gotten its second Meritorious Unit Citation, which he didn't understand, because he was, like most of us, not allowed to see the traffic we collected. He drove an airplane and was not part of ASA proper. When it came to Operations, the Chain of Command was upside down, with the Operations Officer selecting the targets and planning the missions, and the E-5s with the radios in back of our aircraft giving direction in the field. He resented this. It was emasculating. But orders were orders.

His final comment on that award showed his true colors. "Besides, he was only a Zip (Vietnamese)," he said about the victim.

He was an exemplar of White privilege, and only commissioned officers were White. (Commissioned officers who were African-American were also White, magically transformed by their commissions. There were more and more such, and he couldn't afford to piss anyone off. Everyone has friends.) The rest of us were of a lesser breed.

That snobbery existed throughout the Army, and was one of the reasons my father retired. It was why my mother had disliked being 'The Colonel's Lady' so much. They both came from rural poor families. The upper crust officers and their wives annoyed them. Some came from Money and were graceless enough to let that show.

Had this Major not needed his ticket punched with time in a combat zone to make higher rank, I doubt if he would have been in Vietnam at all. Not that he was scared, just the opposite. We weren't on the front lines. He had no bragging rights with his West Point classmates, since no one knew exactly what we did, and we were very careful not to tell them. He was a good soldier in most respects, but being relegated to 'Support' was something that he felt keenly, as a personal failure, I suspect. One thing was sure: awards for his troops would be kept to a minimum. He felt no pride in our accomplishments, which included saving thousands of American and Allied Forces lives by delivering timely intelligence. It lacked the essential glamour of front line combat.

But we had our moments. We had no need to look for trouble. It found us often enough.

My pal Bill, the young MP, was assigned to a supply convoy one day. This was part of our do-the-base-chores, fly-under-the-radar, don't-make-waves, strategy. We contributed some big trucks, and some guys from the motor pool who wanted an adventure went along, as did a couple of others. I think Bill was the only MP, and he went in his usual full 'battle-rattle' of extra weapons and gear. He rode next to a tough, mean, short, and fat Italian-American kid from a big city , who was eager for action, too. This kid strutted around looking for trouble. I suspect he was a bully in high school, and simply didn't comprehend his situation. There were plenty of fights available, and no one needed to pick one. The mortar attacks should have made that plain, but the base was so well protected, and we were so coddled, that some were insulated from reality.

The trip up to Saigon took about three hours because the highway was torn up here and there from war damage. Nothing happened until the return trip, when the trucks were fully loaded with supplies. Then a local-force Viet Cong squad sprang a trap, firing on the trucks from the side of the road. (Since I was not there, I am going on what I can recall of what I was told at the time.) The convoy stopped. Bill was riding next to the tough little Italian-American kid, who was frozen stiff – just sitting there, unable to move.

Bill had to pry his hands off the steering wheel and drag him under the truck, telling him to start firing his weapon at the enemy positions ahead of him. (He never did.) Then Bill ran up and down the road, organizing a defense. Rocket-propelled grenades began to land among the trucks. If one actually hit a truck, then it and the supplies it carried would be lost.

Bill took action. Standing out in the center of the highway, he held his M16 in one hand and his .45 caliber pistol in the other, and directed fire on the enemy, yelling encouragement to the others. As one of the men who saw this said, "He looked like one crazy motherfucker out there." Everyone else fired their weapons, some of them standing and yelling as they did so.

The Viet Cong lost heart, and broke off their attack. No one messes with a crazy man. They ran away, and suddenly it was over. No one badly injured and all of the vehicles intact and undamaged.

By the time they got back to the airfield, the rest of our guys were pumped up. Very excited.

"Don't let anyone tell you that combat isn't a gas!" one of them shouted as they poured into the Flight Operations building to report. The CO listened, and then told them to calm down and go unload the trucks. The little Italian-American kid was very quiet, but no one teased him about his panic attack. Some of the guys had them during mortar attacks. Some were the career soldiers who were also in a combat zone for the first time. One of them was so spooked by the mortar attacks that he moved his bunk into one of the bunkers where he caught pneumonia, and damn near died of that, instead.

Panic was something that was quietly noted, but not discussed. You just hoped that it wouldn't happen again. Everybody got a pass for it – the first time. And most of us never gave in to it. We expected that we would experience some kind of combat at some point, and some, like Bill, looked forward to the day when they could prove their manhood by doing brave deeds.

As the younger men were shouting and slapping each other on the back as if they had just won an important football game, I remember shaking my head. None of them particularly liked Bill, but they were all talking enthusiastically about how he'd led them to victory. They had no idea how lucky they had been. The Viet Cong had not even used a classic 'L' shaped ambush to catch them in a crossfire. Most of the RPGs fell well short of the trucks. And they'd had a good leader, a skinny 19-year-old kid they usually teased for his 'John Wayne' habits. Bill wasn't with them, and when I inquired, I was told he'd had to go to the dispensary for some sort of minor injury.

The next day he came to me at my office.

"You do Awards and Decorations, right?"

I nodded.

"The Medical Officer gave me this, and said I had to turn it in to you."

He handed me a form that I had never seen before. I had to go to our shelf of Army Regulations and look it up.

I said, "Bill, this means you get a Purple Heart."

He was very surprised. "You're kidding! For this? It's a scratch! He took it out with a pair of tweezers. They covered it with a Band-Aid."

He unbuttoned his fatigue shirt and showed me the Band-Aid which was near his sternum above his diaphragm. He'd been wounded, but in a very minor way.

I asked, "What was it?"

"A piece of shrapnel."

"Do you have it?"

"Sure." He dug into his pocket and came out with a tiny glass sample bottle. Inside of it was a wicked-looking, blackened, jagged piece of steel about three-fourths of an inch long.

"Hey, no big deal, right?"

I felt very solemn. "Bill," I said, "I think you should take it. Reading the reg, you have earned it. And if we don't forward this up the chain of command, we'll look like assholes for trying to deny you something you have earned. It will help you with the V.A. after you get out, and give you extra points for a government job. But there's another, more important, reason."

"Yeah?" He looked at me, now very concerned, because I was talking to him like some sort of Dutch uncle. I had never done that before.

"That scratch is right over your aorta. That's the big vessel that carries blood to your heart. If it had gone three inches deeper you would have bled out, and died in about five minutes."

He stared at me for a long time, and turned really pale, as what I said sank in. He sat down hard in the chair next to my desk, suddenly short of breath, his eyes wide. Until that moment, it was all a glorious game. The hard fact that he might actually die in Vietnam hadn't made much of an impression. He got up and bolted out of the room, and that night, for the first time in his life, got very drunk.

He recovered quickly, and appeared for the next guard mount still hauling his extra gear. He was a more sober, less enthusiastic, warrior from that day forward. I went around trying to get affidavits so we could put him up for a Bronze Star with V device for valor, but our new, mean-spirited Commanding Officer had put out the word, and no one would give me one. Bill didn't think he'd done anything special, so he didn't care.

I did the paperwork for his Purple Heart, and walked it into the CO's office myself. He wrinkled his nose as if he smelled something bad, and I said, very quietly, "The other copy of the Medical Officer's form is out there somewhere, and if this is not signed and forwarded, it will make us look bad." He stared at me, a bit surprised at my temerity. Perhaps he'd heard the story about how I'd kept us all from being shamed by killing the phony awards eight months before. Perhaps not, but he could tell I wasn't going to let it go, just because he made an ugly face at me. He reached over and signed it. I took it away rather than leave it there. I didn't want it getting 'lost'.

As I was headed for the office door, he said, "I understand your father is an officer?"

"Yes, sir. Medical Corps, but he retired in May." My way of telling him that he had nothing to fear from that direction, and that it didn't matter. That I wasn't afraid of him.

"So he won't be assigned here?" I could hear the implied sneer in his voice.

"He was here about a year ago, sir. As the Medical Corps IG."

That took him aback. To recover, he said, "That's a really sloppy uniform, Hamit."

His was tight, almost bespoke, because he'd had it tailored.

I shrugged. "This is the way the regs say they are to be worn, sir." Not to mention I'd lost at least ten pounds. "Excuse me, sir. I have to get this in the bag for tomorrow."

One of the other officers later told me that he'd called me a 'barracks lawyer' in their mess. I replied that I took that to mean he thought I was doing my job. But I didn't mean it. I knew his type. He was, in my opinion, a mean, bigoted, ring-knocking son-of-a-bitch: the kind of officer who might lose us this war through his own ambition and cultural blindness.

Bill's Purple Heart arrived about a month later. I watched the ceremony from a distance, and don't know what the new CO said to him, but a week later he was transferred out to a new Radio Research base up North that needed MPs. So I lost the best friend I had there. Nice kid. I missed him.

His was the only Purple Heart awarded to a member of the 156th the entire year I was there. ASA wasn't supposed to bleed. Perhaps that's why he was transferred; for setting a bad example. We had ASA soldiers in other units who received not just Purple Hearts, but awards for valor, up to and including the Silver Star. They didn't plan on that, but sometimes it just turns out that way. The false pride that made this CO deny awards to his subordinates was very inappropriate, and a detriment to morale.

CHAPTER 24

One of our aircraft went down about that time. This really brought the reality of the war home to us. Was the crew alive? Were they hurt?

It was a mechanical failure. The extra gas tanks are in the wings of a U-6. To keep the plane flying level, you have to switch back and forth between them about once an hour. On this particular aircraft, the valve got stuck, or broke, midway between the two tanks, cutting off the fuel to the engine. The motor sputtered to a stop, and the pilot and co-pilot radioed in that they were in trouble, and the plane was going down. They managed to glide it to a safe smooth landing in a rice paddy near a village known to be Viet Cong infested.

The pilot, co-pilot, and operator were all uninjured, but in serious trouble. The received wisdom was that any aircrew down in the Mekong Delta for more than half an hour would likely be captured by the enemy. The operator was at particular hazard. The Viet Cong had a 100,000 piaster bounty out for guys like him.

In Can Tho, we asked for a scramble, and two Huey Cobras went rocketing out toward our downed aircraft to provide cover, followed by a 'slick,' an unarmed Huey with a rescue crew. It was unarmed only in the sense that no weapons were built in. The crew had an M60 .30 caliber belt-fed machine gun they could fire from the big door on the side.

From what was said later, I gathered that it was a near thing for our crew. Viet Cong soldiers were running across the rice paddies toward them as the Cobras swooped in and laid down their patented field-of-fire to discourage them. The crew of the slick got our guys aboard, while adding their own bursts from the M60, and got out of there.

I was sent out in the M151 quarter-ton jeep to pick them up. The pilot and co-pilot got off the helicopter, carrying their weapons and flight bags. Behind them came the operator, an E-5 who was rather short and looked rather pissed off. He had a shock of straw-like blonde hair. He was 18 years old. Headphones crushed his ball cap, and sweat stained his jungle fatigues. He was lugging both of the heavy radios himself, and another jeep, with the Operations officer driving, pulled up behind me. He got in to be taken back to the Operations Building, so the radios and other SIGINT gear could be secured. The pilots, both walking oddly, and looking scared, got in the back of the jeep I was driving, and wanted to go immediately to their hotel, but I was under orders to get them to Flight Operations for a debriefing, and ignored their pleas.

We were all glad to have them back, of course. The entire unit was standing by and waiting. ARDF units did lose planes sometimes, and crews. Sometimes fatally so. It didn't happen often, but was one of those unspoken worries that was always with us every time a plane went on a mission. We did five or six missions every day. You have to be a real fool to shoot at an observation aircraft, but sometimes our birds came back with bullet holes. The pilots and operators routinely sat on their flak vests.

The next day a party of soldiers and mechanics went out to recover the aircraft. It was slightly blown up, with a satchel charge planted by the Viet Cong in the cockpit, and was a total loss. The cockpit and controls took most of the damage, but the repairs could only be done stateside, if it were not scrapped entirely for parts.

A heavy-lift Chinook double rotor helicopter pulled it out, and slowly flew it back the fifty or so miles to Can Tho. I watched as it was gently lowered to the ground, standing next to the Army aircraft mechanic whose responsibility it was. He cursed the Viet Cong, with tears in his eyes. This was his baby, and he took its wounds personally. A few days later, with the wing dismounted and strapped to the top of the fuselage, it was lifted on top of a big flat-bed truck to be driven to Saigon and retrograded to the States. Shortly after that, a replacement craft arrived to continue the mission, already tricked out with the special antennas we used for ARDF.

There was a long delay in operations one day shortly thereafter. A civilian Pilatus PC6 Porter , a short-take-off-and-landing single-wing aircraft, came in carrying a special operations soldier who was going home. He had been in the field a very long time, under miserable conditions, and was so eager to catch his connecting flight that he didn't take heed of the nearly invisible aircraft prop spinning at 3,000 RPM in front of him – walked right into it.

He was killed instantly. The prop hit him three times, tearing his body apart.

There had to be a very slow walk, by a line of soldiers from various units, down the field to find all of his pieces. Otherwise, they might get sucked into an aircraft engine and damage it, or even cause another tragedy. His left arm was found on top of the air traffic control tower. It was the largest chunk aside from his torso and legs. A very sobering reminder of just how deadly a moment's carelessness or inattention could be in the industrial environment of the Vietnam War. You had to pay attention. Relax was not a word we used there.

CHAPTER 25

The airfield commander, a twice passed-over-for-promotion Lieutenant Colonel, was said to be furious at the incident. Not for the loss of a good soldier, but because it upped his accident statistics.

He was eager to impress his superiors with the efficiency and general appearance of 'his' base. The weeds along the sides of the various roads were a particular aggravation for him. Remember what my father said about white rocks appearing every spring on an Army base? He wanted it to look like a stateside base in spring, and expended a great deal of time and effort toward that end.

Having heard of how effective the defoliants we were spraying on the jungle were, he borrowed some from the OV-1 Mohawk company that did some of the spraying, and used the truck normally used for insecticide spraying to douse the whole base. Just once – and it worked a treat. No more weeds. They all died. So did the Kentucky bluegrass patch that the 13th Aviation Battalion installed in front of their Headquarters building. It had been brought in as turf by their transport pilot friends. They replaced it. The new turf also died, and after the third failure, they went to raked sand because nothing, not a single blade of grass, grew in the built-up areas of Can Tho Army Airfield after it was sprayed. The weed problem was solved, and replaced by something much more hazardous: dust. On dry days, just landing an aircraft there became a real adventure. We began training more pilots to fly on Instrument Flight Rules, and not just for tactical reasons.

I have never blamed the airfield commander for exposing me to Agent Orange. He had no idea the harm he did. At the time, the Army and other military branches were assured by the chemical companies that made the stuff that it would not hurt the troops, and this was the received wisdom from on high.

Admiral Zumwalt, whose own son was a Riverine Force officer on the Mekong, authorized its spraying there. His son died young of both Hodgkins and non-Hodgkins lymphoma. Zumwalt later said that, had he known, he would have gone ahead with the spraying anyway, to strip cover from the riverbanks. The Riverine was getting ambushed at every turn, and getting their asses kicked.

A friend of mine, who served there at the time, has legs stained orange from handling the chemicals they sprayed from the PBRs, and he was there for just 58 days. In that short time, he had four boats shot out from under him, and spent the next 18 months in the hospital recovering from his more obvious wounds. This was not the war that everyone saw on televison with the nightly news.

Agent Orange

There were indicators that the defoliants would be more than a short-term problem. I found a single page de-classified report several years ago that described an incident where the barrels used to ship Agent Orange were sold as scrap to a local business.

They were hauled away, and some of the residual compound leaked as they were transported along a road to the scrap yard – and every tree along that road died shortly thereafter! The officer who filed this report was very concerned that the stuff might be stronger than he had been advised.

The Airfield Commander, as you might have already realized, was not the brightest guy around. Had there not been a war on, he would have already been retired at 20 years of service on half-pay. This is where most commissioned officers finish their careers, and there is no shame in it. But he was ambitious.

He took it upon himself to pull an surprise inspection on our Operations Building. Someone had told him (as a joke, I hope) that we were running an illegal casino there. He decided to see for himself. His Sergeant Major, who was supposed to keep him out of trouble, and damn well should have known better, came with him, swagger stick in hand.

A young 19-year-old MP, James C., an E-4 who had already fended off many lookie-loos from other units with a polite refusal, denied them admittance. This building was so restricted that most of us who were in the 156th (and we all had the Top Secret/Crypto security clearance) were not allowed in there. It was not just clearance but 'need-to-know'.

The Airfield Commander, declaring if it was on his base, he could damn well inspect it, moved toward the Quonset hut door, his Sergeant Major on his heels like a faithful dog. This caused the 19-year-old MP to draw and charge his .45. The sound of a round being chambered stopped the Airfield Commander and his CSM in their tracks. They turned, amazed at his effrontery, to see his weapon aimed directly at them.

"Sir, with all due respect," he said evenly, "If you take one more step I am going to blow your asses away."

Admittance required the display of a Worldwide Badge. This was Standard Operating Procedure at every ASA base. It was how every ASA MP was trained, and no one, not even a general officer, could override it. And if he failed, there was another MP behind the door with identical orders, and a 12-gauge shotgun.

The Airfield Commander, trailed by his Sergeant Major – now infuriated – came boiling over across the asphalt road to Flight Operations, where our Company Commander was signing his daily ration of paperwork.

He made a lot of noise, and ranted at our West Pointer for a good 20 minutes. I was in the next office and heard every word. I stuck my head in the door to see if I needed to intervene, my .45 in my hand. Other guys in the offices stopped what they were doing and came into the hallway. The CO might have been a sonofabitch, but he was our sonofabitch.

He didn't even look up at the Airfield Commander, until he stopped to draw breath. He didn't stop signing papers, and spoke very calmly.

"Sir, you were told that that facility is off-limits to you and your staff. You were told that you were not to draw attention to it in any way, by word or deed. That was a direct order from MACV, which you have now violated. The only question left, now, sir, is whether you tell MACV about your breach of our security, or I do."

And he continued signing papers, not bothering to rise or otherwise acknowledge the furious Lt. Colonel standing in front of him.

He and his Sergeant Major left, uttering threats, but two weeks later, they allegedly were both out of the Army, forcibly retired in lieu of a court martial. It takes real talent to be dismissed from the service in the middle of a shooting war.

The young MP was promoted to E-5 (Buck Sergeant). This was an ASA tradition, because such incidents were more common than you might suppose. In Korea, a Major General who attempted a similar 'surprise inspection' was placed in a front-leaning rest (like push-ups but motionless) while the Sergeant of the Guard was called. That took time. Hours.

When I queried James C. about this incident, he barely recalled it. Lots of officers from other units came by and tried to BS their way inside. That was probably because of the casino rumor, or them just having too much time on their hands. The Airfield Commander was the only one to actually try to get to the building door. He was lucky to not be shot, and we were lucky that James C. didn't have to shoot him, because that certainly would have compromised our cover.

The new Airfield Commander never came near the place.

CHAPTER 26

The next month, First Sergeant Pyke noticed that I had not yet been on R&R. R&R was a paid vacation to a big city close to Vietnam. Five days of fun and games and girls – US Government sponsored sex tourism. The girls were whores, of course, because most of the troops did not have the wonderful facilities we had in Can Tho for getting laid. The entire program was arranged and managed by the CIA, I was later told. Your tax dollars at work, and what a proud moment for those agents.

We were, indeed, the most coddled army in history. Destinations included Sydney, Bangkok, Hong Kong, and Kuala Lumpur. The last sounded exotic and interesting, but I was short of money because I had been putting some aside for my 30-day leave, which would be at my expense. It was in savings bonds, which were impossible to simply turn into cash quickly. And besides, I was not some sex-starved grunt in need of relief. Let one of them have the joy of it. But Pyke insisted. Maybe he simply wanted to check that one off his list, or maybe I was getting a little toasty around the edges, but go on R&R I would – and I would damn well like it!

I had to go in to Can Tho and buy some civilian clothes to wear on the trip. No uniforms were permitted. We had to change before getting on the plane in Saigon, and leave our uniforms and field gear in a locker, to be retrieved on our return. It was a perfect opportunity to desert and disappear into Saigon, but there was barbed wire and MPs to prevent that. About 300 guys all boarded the airplane in civvies, each carrying an overnight bag with underwear and toiletries. Some brought bottles of liquor, and those were passed around. When the pilot announced that we had cleared Vietnamese airspace, a ragged cheer went up.

I was not that unhappy to be there. But others were. My seatmate was married to a Vietnamese woman, and had four kids with her. She'd given him a fair ration of hell for making the trip, but he'd been ordered to go, as well, if only to monitor the other men in his unit. He had 'gone native' and contrived to be in Vietnam for eight years, and was the NCOIC for the night convoys from Saigon to Tay Ninh. That was the point of putting everyone in civvies: no rank or Army culture. Just have a good time! Except he'd been appointed to be the hall monitor for this high school field trip. He was a nice guy, one of the good career NCOs that the Army relied upon, who could calm a dispute with a quiet word. He reminded me of SFC Benny Bowman, my favorite DI in Basic. He had that same calm, professional demeanor, and he passed on getting a date for himself, holding up his wedding ring for all to see. That was rare. But some guys, even at the 156th, honored their wedding vows.

"And what do you do?" he asked, more out of politeness than any real interest.

"Me? 71B. Just a clerk-typist."

"Don't say 'just'," he replied. "The Army runs on paperwork. We'd all be screwed without it. That's where the real power lies."

Kuala Lumpur turned out to be a modern city with tall, glass-sided skyscrapers, and looked like Cleveland, were it not for the three-wheeled taxis, the Sikh policemen, wearing turbans and shorts with knee socks, as they directed traffic, and the mix of brown and yellow faces. The same but different.

Before we boarded a bus for the hotel, we had a security briefing from a guy in casual civilian clothes, tall, about 30, with bright red hair. The CIA guy doing his best to blend in, I suppose. There were a lot of American civilians in KL. I wondered what his cover was, or if he even bothered to have one.

"I am not going to ask if any of you are with intelligence or Special Operations," he said, "Because I don't want to know. And I don't want you talking about it. There are about twenty guys in Kuala Lumpur who look and sound just like Americans, and have great stories with which they will try to draw you out. They are actually from the KGB and the GRU, and working for the enemy, and they don't just want spooks. They want anyone who is in administration, transportation or supply. They will offer you money. Don't take it. Some of them are now working for us, too. We will find you and send you home directly from here for trial. For treason."

The room got very quiet when he said that. Everyone looked at each other and then back at him. Satisfied he had made his point, he added, "Don't talk to the girls you meet here, either, about what you do. They're all Malaysian Chinese, and most of them speak better English than you do. But some of them also get a paycheck from the Chinese Secret Service. Don't tell them anything, and for God's sake, don't fall in love with them. They're very charming, but they're whores, and not to be trusted. That said, I hope you have a great time. Now get out of here."

And I did have a great time. It was like being back in high school again. We went in groups in hired cars to the beach and bowling and the movies (first-run films from the USA) and, while the sex was not as good as in Ben Xe Moi, no one was shooting at us either. There was one notable young lady named Ginny, who was not all that pretty, but made up for it in sheer spirit and good humor. She laughed when she told us about her former 'boyfriends', some of whom sent letters proposing marriage, and gave us an insight into her side of the business. She had a new one every week, and had trouble keeping them straight in her mind after two years. That was 100 American guys she'd slept with, at a hundred dollars each plus tips. (The girls' fee was $20 per day for 24 hours a day.) She must have been very, very good at her job. She looked sweet and innocent but was anything but.

"We have a school for this: teaching how to make love," she said in an unguarded moment. "I am an instructor in that school." She was 20 years old, and I half wanted to marry her myself, even if she was someone else's date.

But I was also a little paranoid. When a Chinese stripper stuck her boob in my face as she was performing for the entire group, I felt no excitement, but froze, feeling like I might be on "Candid Camera". I felt over-exposed, if not embarrassed – and wary, because I could tell her heart wasn't in it. I gave her five dollars and she moved on to the next guy, never cracking a smile. He, too, got a boob in the face.

Truth to tell, I missed the girl in Ben Xe Moi. It wasn't all about the money with her. She liked me. And I liked her enough to keep it exclusive for the last few months I was there.

I went back to Can Tho and had a serious talk with First Sergeant Pyke about extending my time in Can Tho. It wasn't just the extra pay for being in a combat zone, and the cheap sex and booze. I also wanted that extra stripe to E-5. Before I'd left, I'd had to fill out a preference sheet for my next assignment. I really wanted Japan, and we had a big field station at Chitose. My father had been stationed at Camp Zuma during the Occupation. Our house was filled with Japanese art and other souvenirs. I'd absorbed some of the culture growing up. But you never get your first choice, so I tried to game the system by choosing Frankfurt first, and Bavaria second, and Japan third. My orders were in, and it was not just Frankfurt but the European headquarters for ASA. A really great assignment that I probably didn't deserve. That was the luck of the draw.

Pyke told me it was a dream assignment. One he'd never been able to get himself despite more than 20 years in ASA. He had kids in high school and thought they would benefit from living in West Germany. If I extended, there was no guarantee I would get orders for Frankfurt again. I'd most likely draw a stateside assignment, with all its spit and polish and neverending BS, for being such an ingrate as to turn Frankfurt down. In fact, he'd make sure of it.

So I didn't file for an extension, and took the orders for Frankfurt. I can take a hint.

But before I went, the war began to heat up. I was 'short', counting down my last 100 days in country, and saying good bye to the closest thing I had to a girlfriend. I often think of her kindly, but it was never going to be more than it was. She had a husband who was missing in action a long time. She never said which army he was with, and I never asked.

The winter weather gave the Viet Cong added cover for an attack, and they began to probe the airfield. One night a very skilled Huey Cobra pilot pinned one against the chain link fence he was climbing over. He radioed TOC (theTactical Operations Center), "Come get this guy before I have to blow him away ."

That was done very quickly. The minigun would have spread his guts like strawberry jam, and no one wanted to clean up the mess. This sapper was later Chu Hoied (persuaded to defect) and demonstrated his skills for a group of MI officers. He stripped down naked, and then went through a triple strand of concertina barbed wire, like those we had on the perimeter in front of the Claymore mines. This was to keep the sappers from stealing them or just turning them around so they would fire at us instead. It took him less than five minutes, and he suffered not a single scratch. Impressive and very scary.

CHAPTER 27

Christmas had been insane with everyone on the base firing their weapons into the air for several hours. What goes up must come down; there was a rain of bullets all over the base and in the surrounding civilian areas. Those of us still sober took refuge in the bunkers. Just before New Year's, when the whole sorry performance would be repeated, we had a full Alert. Not just for those of us on our assigned bit of the perimeter, but everyone. 'Get your gear and get ready to move out!' was the order.

We were all in the barracks, donning flak jackets, helmets, and checking our M16s and ammunition, and making sure we had two canteens of water and a day's rations. Everyone was very quiet, but no one was panicking, that I could see. Most of the ASA guys had been through the Tactical Training Course at Fort Devens and knew what to do. We were calm and confident, but not eager. Walking around, I could see everyone had a quiet resolve, an attitude that we were soldiers first.

Except for my NCOIC, who was sitting on a bunk, looking scared as hell and staring straight ahead. He had less time left than I did. He would be going home in less than a month. And he was in what the shrinks call a 'disassociative state'. Someone waved their hand in front of his eyes and got no reaction. He was supposed to be our platoon leader during the coming attack. Finally, someone took his M16 from his hands and cleared it. Still no reaction. He was frozen in place, and not present.

We looked at each other, and then, with hand signals, made our own assignments for squad and fire team leaders. The matter had never come up, and we had no ops plan to consult. None of the MPs were with us. We just figured it out for ourselves.

Then the Alert was cancelled as the 'all clear' siren sounded. And it took that E-7 several minutes to come back to himself. No one ever said anything about this breakdown. He was career, and two years away from his pension. He was also 'short' and about to go away. Why make trouble for him, prick though he was?

The attack never came that night. The Viet Cong weren't quite ready. About two weeks later, on January 13, 1969, they were, and attacked in force. It made the New York Times and other papers 'back in the world'.

My father sent me a letter, urging me once more not to extend. I'd already decided not to. Europe awaited, and the Viet Cong did major damage, killing eight of our guys and blowing up two large double rotor heavy-lift Chinook helicopters. They had gone for the Cobras and settled for those.

The sappers cut the throats of three newbies assigned together in one perimeter bunker at the far end of the runway. They'd all fallen asleep. This was a big mistake on someone's part, because they hadn't been in-country long enough to take the Viet Cong seriously.

Some of the Cobra pilots were killed running to the new revetments to scramble their aircraft. The rotary wings raised a huge cloud of dust that obscured vision. It was rumored that the Viet Cong were driving a captured quarter-ton down the runway and killing anyone they met, while setting more satchel charges. The fight went on all night.

We'd known an attack was coming, and tripled the guard on our part of the perimeter; the location had been changed. Rather than two to a bunker on the far part of the berm next to the rice paddies, we were now six to a bunker (newly built that day and only waist high with no roof) close to the Main Gate, and directly across from the PA&E (Pacific Architects and Engineers) compound about a mile away across the rice paddies.

This was filled with bulldozers and other heavy equipment, and guarded by Nung tribesmen – civilian mercenaries. We could see the flares popping overhead and the mortars hitting the yard. It was oddly beautiful. The VC never came near us. No one messed with the Nungs. They were known to cut heads off.

It was a disappointment for some of us. After months of constant mortar attacks, we really wanted a piece of those VC bastards. The Newbies all thought we were crazy when we said this. Maybe we were.

When we got into position before dark, I climbed up on top of the wall of sandbags and laid down to get some sleep. With so many of us there, this was a perfectly safe thing to do until the attack started.

Flight line, Can Tho Army Airfield

When the fireworks started at PA&E, someone shook me awake. I rolled off, hitched my rifle on my shoulder and walked back a few paces, unbuttoned myself and delivered an impressive stream of urine into a pile of sand. I didn't want to fight with a full bladder. I turned back, walked to the sandbag wall, pulled the charging level on my M16, and chambered a round, looking out as if to say: "Where are they?"

I felt perfectly calm. I had envisioned moments like this many times growing up. It was the family business. I am a soldier. My attitude steadied some of the new guys. We'd already picked out fields of fire. Other guys thought it weird, but they hadn't been there long, and it was their first big attack. Mortars were nothing next to this. The guys who'd been there for the Tet Offensive in 1968 were all long gone. That was another downside of the one-year tour. Institutional memory was quickly lost.

The enemy never came our way that night. We'd been moved to keep us out of the fight. We were too valuable an asset to risk defending against a low-level ground attack. Had the Viet Cong known who we were, they would have gone for our aircraft and personnel, and not the Cobras. We came through it all without a scratch. Nothing to brag about, but nothing to be ashamed of, either. We did our job, which was to do nothing and stay out of the way.

I was given a recommendation to promotion in my 201 file, but it was a gesture. It would have no effect at my next post. The Army Commendation Medal I was recommended for would be a help, since it would generate points for all future promotion boards.

The closer you got to leaving, the more nervous you became. It was called 'short-timer's syndrome' – you became hyperaware of your surroundings, and this vigilance would carry over into my future life and eventually manifest, years later, as PTSD.

I bid a fond farewell to my girlfriend, and stopped going to Ben Xe Moi. This was also a health measure. You didn't want to carry certain diseases home with you, and the Army was selectively blind about those who contracted them. Under the UCMJ you could be court-martialed for getting one, but we got shots instead. You can't court martial an entire army.

At the end of my year there, I found myself in jungle fatigues, without a weapon, waiting in a holding barracks at Tan Son Nhut, and sharing a bottle of Jack Daniels with a tall, lean paratrooper who looked like he had American Indian heritage, and turned out to be from Kentucky.

He'd just served a year in the 101st Airborne. He looked sad. He shared with me what the 'real' war had been like, as we passed the bottle back and forth. He did not call me a 'REMF' either.

We were both avoiding a braggart going home from the 525th Military Intelligence Group. One of those 'I've got a secret' types, who confided that he'd been involved in something called the Phoenix Program. Not directly, which was probably a good thing, since he was as fat as he was tall, and never left Saigon. I thought the confidences he delivered about a secret assassination program pure bullshit. It didn't seem like anything he would actually be involved in, or even know about, because he was remarkably indiscreet for a man who'd spent a year in-country in Military Intelligence. We were all strangers to each other, and if anyone was a REMF, it was him. It was not until more than a year later that the Phoenix Program became the subject of headline news. At which point, I cheerfully reported this blabbermouth for disclosing classified information. My excuse for not doing so sooner was that I didn't think that any of it was true. It sounded to me like bad fiction, and counter-intuitive to 'winning hearts and minds'. Our government wasn't supposed to do such things. Were they? Weren't we the good guys?

I was more interested in the paratrooper, who was humble and modest. He'd earned medals for valor, which he was too embarrassed to talk about. They were no substitute for the friends he'd lost. I envied him his opportunities, and said so. He replied that not everyone could be in direct combat. Or should be.

"You had your job, and I had mine," he said, taking another swig.

"Yours was harder."

"It's what I signed up for," he said; "Kill a commie for Christ, and all that shit. God, I was so naive! I didn't think I'd spend my time taking a hill and then giving it back, or watching my buddies getting killed every day. You stop making friends with people after awhile. And the brass are clueless about how it feels. You're sitting there with the body bags filled with your friends, and they send out a helicopter with ice cream to make you feel better. Ice cream. Fucking ice cream, like we're little children, and that will make it all right!"

He looked away, silent for a long moment. "You've got to respect the guys on the other side. They're very tough. Great soldiers." He took another drink, and said very quietly, "Don't tell anyone I said so, but we got our asses kicked at the A Shau Valley."

It was at that moment that I knew that the war would be lost.

CHAPTER 28

The entire inside of the airplane exploded with cheers when the pilot came on the intercom and announced that we had cleared the territorial limits of South Vietnam.

Most of us were a little bit drunk. The rule about no liquor in the holding area had been observed in the breach. The jetliner belonged to a military contractor, and the stewardesses were all American! Roundeyes! Outside of entertainers passing through on a USO tour, we had not seen their like in a year. All about twenty years old, smiling and very pretty, dressed in light blue uniforms, with blonde hair and red fingernails, just like the girls that many had left behind.

Once the noise settled down, they were very efficient in handing out blankets and pillows so everyone could start sleeping it off. When meal time came, they delivered the trays and beverages with smiles, kind words, and polite refusals for their phone numbers.

It was supposed to be a direct flight to Hawaii, and then on to Oakland. The officers were segregated in the front, and relied upon the few senior NCOs to control what was going on in the back. And it was a happy party, up until the moment one of the engines flamed out, and we had to divert to Okinawa.

The mood became somber as we all contemplated the possible irony of surviving a year in the war zone, and perhaps dying when that over-aged rickety piece of junk we were on crashed into the Pacific. Speed was reduced to save fuel and we limped into a civilian airfield on Okinawa after dark. The stewardesses, no longer smiling, went about reassuring every one that there was no real problem. They were wrong.

Once on the ground, we were told that we would not be leaving the airplane. The local Japanese authorities did not want 300 crazed Vietnam veterans loose on their streets, loaded with money, looking for love or booze, or both, and annoying the locals.

This was not exaggerated fears on their part, but the bitter fruit of past experience. American Military Police were en route to control us, and we were all part of a military formation under transit orders. Our 'unit commander', a lieutenant colonel, went to the cockpit to reason with the people who didn't want to let us off the plane. The engine repairs, even at express speeds, would take all night, and the internal temperature in the aircraft was rising, and the air getting foul.

He could be heard in the back of the plane making his points in rich, colorful language that told us that he'd come up through the ranks and had NCO experience. It was reassuring.

Finally, a compromise was reached. The troops would be bused to a nearby hotel, which would be in lock down, and secured by the Military Police so none of them could leave. No rooms were available, but there would be a bar and restaurant and a steam bath.

The aircraft crew would be taken to another hotel so they could rest. The stewardesses were visibly relieved to hear this. Their charm-school smiles had begun to slip a bit. Being trapped all night in that aluminum can with 300 Vietnam veterans was beyond the call of duty.

We were allowed to deplane, thirty at a time, counted off and our names checked off a list, to board buses driven by local drivers, who pretended to speak no English. There were two MPs on each bus. The hotel was a few minutes away. Once there, we filed off the bus, and formed up to 'get the word' from one of the senior NCOs.

We were not to leave the hotel, and would be arrested by the MPs if we tried. There were no rooms available. Sleeping would be done on the floor, if at all. The good news was that the company that owned the airliner was bringing in real American food, the bar was open, and there was a steam bath! And when we got inside, there were hamburgers, chips, soda, and beer for free, or you could buy booze at the bar for prices that both shocked us and limited consumption.

The best of both words, I thought. I wanted the stink of Vietnam off of me and lined up for the steam bath first. You could strip down, of course, but the Okinawan girls wore modest swim suits and were not selling anything but cleanliness. They were all attractive, but very strong, coming from families that did deep-ocean fishing. They were very firm about that no-sex rule, and ready to punch out anyone who didn't see it their way.

The other guys in the room policed the situation. What if she was your sister? one of them asked. Not everyone was all that drunk. Vietnam was behind us, and so was easy sex on the cheap.

I must have been very drunk. I felt so relieved to be going home! I went back four times for steam baths, and the steam bath girls were weary and unsmiling by the time they closed it down. The tips were generous and they walked away with about a thousand dollars each, but you could tell that they did not have a good time. The food was gone about midnight, and the bar closed; there was nothing to do but to try to get some sleep, or read.

I read and tried to sober up. I had leave coming, and no intention of spending it all with my parents. I wanted to go back to Iowa City for a few days and see friends, if I had any left. I also wanted to do some writing and get back into doing some photography. My parents were not in Chicago proper, but the suburb of Deerfield. I had never been to Chicago, but I'm a big city boy and was looking forward to exploring it.

The plane ride back to the military airport near Oakland, California was uneventful. The stewardesses, refreshed now, were their old smiling selves. The plane was fixed and reloaded, and the pilot and co-pilot relieved by a fresh crew. The 300 of us settled in and slept most of the way home, awakened by the pilot's voice telling us to look out the windows and see the Golden Gate Bridge. The officers and senior NCOs looked very neat and clean. Apparently, rooms with showers had been found for them at the aircrew hotel. No one was surprised by this, or resented it. It was the U.S. military, and Rank Had Its Privileges.

We were deplaned and processed. You carried your 201 (personnel) file with you, so that you and it would arrive at your next duty station together. But it was sealed so you couldn't look at it or tamper with it. You had to present your orders and surrender it for inspection before you could get your plane tickets home and vacation pay advance. I was issued a new Class A dress green uniform, and had to turn in my jungle fatigues. I had lost 20 pounds, and it was simpler to get a new uniform than to have the old one tailored. I had two patches and my service ribbons put on. One patch was the gaudy red and yellow phallic one for MACV (Military Assistance Command, Vietnam) and the other was the infamous 'Lightning-fast chicken-fucker' with an eagle clutching lightning bolts on a dark blue background for the Army Security Agency.

"I thought there was no ASA in Vietnam," said a bemused Air Force payroll clerk as she counted out my vacation pay to the penny.

"There isn't," I assured her, practicing the agreed-upon lie. "I'm going to ASA Europe next." She nodded and motioned to the next guy in line.

Wearing your Class As got you an automatic discount with the airlines, who were trying to show their patriotism. It got you wary looks from civilians. The drumbeat of anti-war propaganda had not yet created those disgraceful episodes where people would show up at the base or the airport to demonstrate against returning soldiers, and even spit upon them, but being a soldier was no longer the patriotic and honorable profession it had been when our fathers were in the service. I wore my uniform with pride, even when I traveled back to Iowa City for a few days to see friends, who were suddenly shy and embarrassed to see me. I wondered if it was the uniform or the undercover work that made them draw back. But no one brought that up, nor did anyone call me a 'baby killer.' None of my old girlfriends were around. I felt at loose ends and didn't linger.

I purchased a new Ampex cassette tape recorder and had the idea of writing an article for publication to make a few bucks. A friend was involved with a company that had a high-tech product for exploring the ocean (in Iowa!). I worked up an article about it and mailed it to Popular Mechanics.

This shows you how naive I was then. The first rule of freelancing is that you get the assignment first, and then do the article, for an editor who has agreed to read it. I didn't know that, or that such a submission was considered little better than a press release. I did get my first printed rejection slip. Progress of a sort. I was in the game and trying.

Back in Deerfield, my 'hold baggage' arrived in a wooden crate which, when my father and I opened it in the garage, turned out to be crawling with six-legged immigrants from Vietnam. A can of bug spray soon took care of that problem, but my father wondered why I had bothered. There were a few books (including my USAFI College Algebra textbook), the civilian clothes I had worn in Kuala Lumpur, and my jungle boots. Everything else had to be thrown away.

My father, as he always had been, was fully engaged in work. The new job did not suit him, and he was already looking for another, but, in the meantime, would do his best and soldier on. Now that I was 24, he would invite me to have a drink with him when he came home at night. But otherwise he was too busy. My mother, as always, was smiling and trying to make the best of the situation, but hated Chicago. It was too cold, and she had no friends there.

Before I left, my parents presented me with an expensive lined brown leather overcoat. It was cold in Germany, they had heard. I really liked the coat, and said so. They bought it at Marshall Field's, but it had been made in Germany. They were both anxious about me. Dad was quietly amazed that I had survived two years in the Army without even an Article 15, much less a court-martial.

"What is that you do?" he asked one night.

"Type, mostly. I'm also a courier."

"Your outfit? Radio Research? What is that?"

"Military Intelligence. And I can't talk about it. My work is so secret even I don't know what I'm doing."

That got a laugh. He got the point, and didn't ask again. "My son, the spy," he said sarcastically. "When you get over there, try to get a real job, or go to OCS and become an officer."

"OCS is for combat arms," I replied. "If I do go that way, a direct commission is more likely."

"Not unless you finish college. The Army is going to require all officers to have a bachelor's degree soon. I read that in Army Times." Despite being retired, he still subscribed to that weekly to keep track of old friends. He really did miss it. My mother did not, so that was the end of that. There was no going back. Being a General's spouse requires executive skills she'd never had the chance to acquire.

I was still in, with almost two years to go, as ill-advised as he thought that might be, going to a plush assignment like the one he'd always wanted. Europe. I could tell he was envious and surprised.

Vietnam was behind me, and I had come through that without a scratch. The biggest risk ahead of me was paper cuts, I thought. If I kept my nose clean and did my job, it might be a career after all. ASA was not the real Army, and I had lucked out by getting into it. My training troubles were history, and I had done superior work in Vietnam. The pay was lousy, but I could make rank and improve that. I knew how to work the system. I thought I finally had it made.

-TO BE CONTINUED-

LIST OF ILLUSTRATIONS

ACKNOWLEDGMENTS

My father, COL. Harold F. Hamit, MD, FACS
Bill Krueger
Gwen Eloise Cortez Kaneshiro
David Yano
Jerry Orlemann
Ezio Tozzi
Jim Price

and

the military historians at Fort Huachuca, AZ
Gretchen Wayne, of BatJac Productions
Fort Devens Historical Museum

Made in United States
Cleveland, OH
26 September 2026

44836485R00125